DIFFERENTIATION RULES

General Formulas

Assume u and v are differentiable functions of x.

Constant: $\dfrac{d}{dx}(c) = 0$

Sum: $\dfrac{d}{dx}(u + v) = \dfrac{du}{dx} + \dfrac{dv}{dx}$

Difference: $\dfrac{d}{dx}(u - v) = \dfrac{du}{dx} - \dfrac{dv}{dx}$

Constant Multiple: $\dfrac{d}{dx}(cu) = c\dfrac{du}{dx}$

Product: $\dfrac{d}{dx}(uv) = u\dfrac{dv}{dx} + v\dfrac{du}{dx}$

Quotient: $\dfrac{d}{dx}\left(\dfrac{u}{v}\right) = \dfrac{v\dfrac{du}{dx} - u\dfrac{dv}{dx}}{v^2}$

Power: $\dfrac{d}{dx}x^n = nx^{n-1}$

Chain Rule: $\dfrac{d}{dx}(f(g(x)) = f'(g(x)) \cdot g'(x)$

Trigonometric Functions

$\dfrac{d}{dx}(\sin x) = \cos x$ $\qquad \dfrac{d}{dx}(\cos x) = -\sin x$

$\dfrac{d}{dx}(\tan x) = \sec^2 x$ $\qquad \dfrac{d}{dx}(\sec x) = \sec x \tan x$

$\dfrac{d}{dx}(\cot x) = -\csc^2 x$ $\qquad \dfrac{d}{dx}(\csc x) = -\csc x \cot x$

Exponential and Logarithmic Functions

$\dfrac{d}{dx}e^x = e^x$ $\qquad \dfrac{d}{dx}\ln x = \dfrac{1}{x}$

$\dfrac{d}{dx}a^x = a^x \ln a$ $\qquad \dfrac{d}{dx}(\log_a x) = \dfrac{1}{x \ln a}$

Inverse Trigonometric Functions

$\dfrac{d}{dx}(\sin^{-1} x) = \dfrac{1}{\sqrt{1 - x^2}}$ $\qquad \dfrac{d}{dx}(\cos^{-1} x) = -\dfrac{1}{\sqrt{1 - x^2}}$

$\dfrac{d}{dx}(\tan^{-1} x) = \dfrac{1}{1 + x^2}$ $\qquad \dfrac{d}{dx}(\sec^{-1} x) = \dfrac{1}{|x|\sqrt{x^2 - 1}}$

$\dfrac{d}{dx}(\cot^{-1} x) = -\dfrac{1}{1 + x^2}$ $\qquad \dfrac{d}{dx}(\csc^{-1} x) = -\dfrac{1}{|x|\sqrt{x^2 - 1}}$

Hyperbolic Functions

$\dfrac{d}{dx}(\sinh x) = \cosh x$ $\qquad \dfrac{d}{dx}(\cosh x) = \sinh x$

$\dfrac{d}{dx}(\tanh x) = \operatorname{sech}^2 x$ $\qquad \dfrac{d}{dx}(\operatorname{sech} x) = -\operatorname{sech} x \tanh x$

$\dfrac{d}{dx}(\coth x) = -\operatorname{csch}^2 x$ $\qquad \dfrac{d}{dx}(\operatorname{csch} x) = -\operatorname{csch} x \coth x$

Inverse Hyperbolic Functions

$\dfrac{d}{dx}(\sinh^{-1} x) = \dfrac{1}{\sqrt{1 + x^2}}$ $\qquad \dfrac{d}{dx}(\cosh^{-1} x) = \dfrac{1}{\sqrt{x^2 - 1}}$

$\dfrac{d}{dx}(\tanh^{-1} x) = \dfrac{1}{1 - x^2}$ $\qquad \dfrac{d}{dx}(\operatorname{sech}^{-1} x) = -\dfrac{1}{x\sqrt{1 - x^2}}$

$\dfrac{d}{dx}(\coth^{-1} x) = \dfrac{1}{1 - x^2}$ $\qquad \dfrac{d}{dx}(\operatorname{csch}^{-1} x) = -\dfrac{1}{|x|\sqrt{1 + x^2}}$

Parametric Equations

If $x = f(t)$ and $y = g(t)$ are differentiable, then

$$y' = \frac{dy}{dx} = \frac{dy/dt}{dx/dt} \quad \text{and} \quad \frac{d^2y}{dx^2} = \frac{dy'/dt}{dx/dt}$$

THOMAS'
CALCULUS

ELEVENTH EDITION

PART TWO

THOMAS'
CALCULUS

ELEVENTH EDITION

PART TWO

Based on the original work by
George B. Thomas, Jr.
Massachusetts Institute of Technology

as revised by

Maurice D. Weir
Naval Postgraduate School

Joel Hass
University of California, Davis

Frank R. Giordano
Naval Postgraduate School

PEARSON
Addison
Wesley

Boston San Francisco New York
London Toronto Sydney Tokyo Singapore Madrid
Mexico City Munich Paris Cape Town Hong Kong Montreal

Publisher:	Greg Tobin
Acquisitions Editor:	William Hoffman
Managing Editor:	Karen Wernholm
Senior Project Editor:	Rachel S. Reeve
Editorial Assistants:	Mary Reynolds, Emily Portwood
Production Supervisor:	Julie LaChance James
Marketing Manager:	Phyllis Hubbard
Marketing Assistant:	Heather Peck
Senior Manufacturing Buyer:	Evelyn Beaton
Senior Prepress Supervisor:	Caroline Fell
Associate Media Producer:	Sara Anderson
Software Editors:	David Malone, Bob Carroll
Senior Author Support/ Technology Specialist:	Joe Vetere
Supplements Production Supervisor:	Sheila Spinney
Composition and Production Services:	Nesbitt Graphics, Inc.
Illustrations:	Techsetters, Inc.
Senior Designer:	Barbara T. Atkinson
Interior Design:	Geri Davis/The Davis Group, Inc.
Cover Design:	Barbara T. Atkinson
Cover Photograph:	© Benjamin Mendlowitz

Dedicated to

Ross Lee Finney III

(1933–2000)

Scholar, Educator, Author,

Humanitarian, Friend to all

For permission to use copyrighted material, grateful acknowledgment is made to the copyright holders on page C-1, which is hereby made part of this copyright page.

Many of the designations used by manufacturers and sellers to distinguish their products are claimed as trademarks. Where those designations appear in this book, and Addison-Wesley was aware of a trademark claim, the designations have been printed in initial caps or all caps.

Library of Congress Cataloging-in-Publication Data
Weir, Maurice D.
Thomas' calculus.—11th ed./based on the original work by George B. Thomas, Jr., as revised by Maurice D. Weir, Joel Hass, Frank R. Giordano.
    p. cm
   Rev. ed. of: Thomas' calculus. 10th ed./ ...as revised by Ross L. Finney, Maurice D. Weir, and Frank R. Giordano. 2000.
   ISBN 0-321-22651-8 (alk. Paper)
   1. Calculus. I. Title: Calculus. II. Hass, Joel. III. Giordano, Frank R. IV. Finney, Ross L. Thomas' calculus. V. Title.

QA303.2. W45 2004
515–dc22   2003063270

1 2 3 4 5 6 7 8 9 10-QWT-07 06 05 04

CONTENTS

16 Integration in Vector Fields 1143

Appendices AP-1

Answers A-1

Index I-1

A Brief Table of Integrals T-1

Credits C-1

PREFACE

OVERVIEW In preparing the eleventh edition of *Thomas' Calculus*, we have worked to capture the style and strengths of earlier editions. Our goal has been to revisit the best features of the *Thomas' Calculus* classic editions while listening carefully to the suggestions of our many users and reviewers. With these high standards in mind, we have reconstructed the exercises and clarified some difficult topics. In the words of George Thomas, "(We) have tried to write the book as clearly and precisely as is possible." In addition, we have restructured the contents to be more logical and in alignment with the standard syllabus. In looking backward, we have learned much to help us create a useful and appealing calculus text for the next generation of engineers and scientists.

In the eleventh edition the text introduces students not just to the methods and applications of calculus, but also to a mathematical way of thinking. From the exercises to the examples to the narrative that develops the concepts and reveals the theory in readable language, this book is about thinking and communicating mathematical ideas. Calculus contains many of the key paradigms of mathematics, and it marks the real beginnings of how to think about physical and mathematical subjects in a precise and logical way. We try to help students achieve the mathematical maturity required to master the material and apply its power. The insights that come from a deep understanding are well worth the effort.

After completing this book, students should be well versed in the mathematical language needed for applying the concepts of calculus to numerous applications in science and engineering. They should also be well prepared for courses in differential equations, linear algebra, or advanced calculus.

Changes for the Eleventh Edition

EXERCISES Exercises and examples play a crucial role in learning calculus. We have included in this new edition many of the exercises that appeared in previous editions of *Thomas' Calculus*, and which constituted a great strength of those editions. Within each section we have organized and grouped the exercises by topic, progressing from computational problems to applied and theoretical problems. This arrangement gives students the opportunity to develop skills in using the methods of calculus and to deepen their appreciation and understanding of its applications and coherent mathematical structure.

RIGOR The level of rigor, while comparable to earlier editions, is more consistent throughout. We give both formal and informal discussions, making clear the distinction between the two, and we include precise definitions and accessible proofs for the students.

The text is organized so the material can be covered informally, giving the instructor a degree of flexibility. For example, while we do not prove that a continuous function on a closed and bounded interval has a maximum there, we do state this theorem carefully and use it to prove several subsequent results. Moreover, the chapter on limits has been substantially reorganized, with greater attention to both clarity and precision. As in previous editions, the limit concept is still motivated by the important idea of obtaining the slope of the line tangent to a curve at a point on it.

CONTENT   During the preparation of this edition, we have paid considerable attention to the suggestions and comments from users of previous *Thomas' Calculus* editions and from our reviewers. This has resulted in extensive revisions and changes to several chapters.

- **Preliminaries**   We have rewritten Chapter 1 as a brief review of the elementary functions. While many instructors may choose to skip the chapter, it allows for easy reference and review by the student, standardizes notation, and indicates what is assumed as background material. It also contains some helpful material that many students may not have seen, such as the pitfalls of relying entirely on a calculator or computer to give the graph of a function.
- **Limits**   Included in Chapter 2 are epsilon-delta definitions, proofs of many theorems, limits at infinity and infinite limits (and their relationship to asymptotes of a graph).
- **Antiderivatives**   We present the derivative and its important applications in Chapters 3 and 4, concluding with the antiderivative concept, which sets the stage for integration.
- **Integration**   After discussing several examples of finite sums, we introduce in Chapter 5 the definite integral in its traditional setting of the area under a curve. Following the treatment of the Fundamental Theorem of Calculus, bridging derivatives and antiderivatives, we present the indefinite integral, along with the Substitution Rule for integration. The traditional chapter on applications of definite integrals follows.
- **Techniques of integration**   The main techniques of integration, including numerical integration, are presented in Chapter 8. This follows the introduction of the transcendental functions, where we define the natural logarithm as an integral and the exponential function as its inverse.
- **Differential equations**   The bulk of the material on solving basic differential equations is now organized into a single Chapter 9. This organization allows for greater instructor flexibility in the coverage of those topics.
- **Conics**   At the request of many users, Chapter 10 on the conic sections has been fully restored. This chapter also completes the material on parametric equations by giving parametrizations of parabolas, hyperbolas, and cycloids.
- **Series**   In Chapter 11 we have restored the more complete development of the series' convergence tests that appeared in the ninth edition. We also include a brief section introducing Fourier series (which may be omitted) at the end of the chapter.
- **Vectors**   To avoid repetition of the central algebraic and geometric ideas, we have combined the treatment of two- and three-dimensional vectors into a single Chapter 12. This presentation is followed by a chapter on vector-valued functions in the plane and in space.
- **The real numbers**   We have written a brief new appendix on the theory of real numbers as it applies to calculus.

ART We realize that figures and illustrations are a critical component to learning calculus, so we have taken a fresh look at all of the figures in the book. When revising existing figures and creating new ones, we worked to improve the clarity with which the figures illustrate their associated concepts. This is especially evident with the three-dimensional graphics, where we were able to better indicate depth, layering, and rotation (see figures below). We also attempted to ensure a consistent and pedagogical use of color and assembled a team dedicated to proofreading the completed pieces.

FIGURE 6.11, page 402

Finding the volume of the solid generated by revolving the region (a) about the y-axis.

$x = \dfrac{2}{y}$

$R(y) = \dfrac{2}{y}$

(a)

$x = \dfrac{2}{y}$

$\left(\dfrac{2}{y}, y\right)$

$R(y) = \dfrac{2}{y}$

(b)

FIGURE 6.13, page 403

The cross-sections of the solid of revolution generated here are washers, not disks.

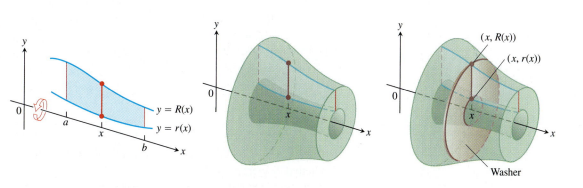

$y = R(x)$

$y = r(x)$

$(x, R(x))$

$(x, r(x))$

Washer

Continuing Features

END-OF-CHAPTER REVIEWS AND PROJECTS   In addition to problems appearing after each section, each chapter culminates with review questions, practice exercises covering the entire chapter, and a series of Additional and Advanced Exercises serving to include more challenging or synthesizing problems. Most chapters also include descriptions of several student projects that can be worked on by individual students, or groups of students, over a longer period of time. These projects require the use of a computer and additional material that is available over the Internet at **www.aw-bc.com/thomas**.

WRITING EXERCISES   Writing exercises placed throughout the text ask students to explore and explain a variety of calculus concepts and applications. In addition, each chapter end contains a list of questions for students to review and summarize what they have learned. Many of these exercises make good writing assignments.

ANSWERS   Answers are provided for all odd-numbered exercises when appropriate, and these have been carefully checked for correctness.

MATHEMATICAL CORRECTNESS   As in previous editions, we have been careful to say only what is true and mathematically sound. Every definition, theorem, corollary, and proof has been reviewed for clarity and mathematical correctness.

WRITING AND APPLICATIONS   As always, this text continues to be easy to read, conversational, and mathematically rich. Each new topic is motivated by clear, easy-to-understand examples and is then reinforced by its application to real-world problems of immediate interest to students. A hallmark of this book has been the application of calculus to science and engineering. These applied problems have been updated, improved, and extended continually over the last several editions.

TECHNOLOGY   In a course using the text, technology can be incorporated according to the taste of the instructor. Each section contains exercises requiring the use of technology; these are marked with a **T** if suitable for calculator or computer usage or are labeled **Computer Explorations** if a computer algebra system (CAS, such as *Maple* or *Mathematica*) is required. While we continue to provide support for technology, we have toned down its visibility within the chapters from the tenth edition.

EARLY TRANSCENDENTALS   For instructors who require an earlier treatment of the calculus of transcendental functions, we have prepared an *Early Transcendentals* version of this text, in which the exponential and logarithmic functions are introduced in the first chapter. Their limits, derivatives, and integrals are given in Chapters 2 through 5, along with those for polynomials and other algebraic functions. Examples and exercises involving the transcendental functions are then interlaced throughout those chapters as the various calculus topics are developed.

Text Versions

STUDENT EDITION OF THOMAS' CALCULUS, Eleventh Edition
Complete (Chapters 1–16), ISBN 0-321-18558-7
Part One, Single Variable Calculus (Chapters 1–11), ISBN 0-321-22642-9
Part Two, Multivariable Calculus (Chapters 11–16), ISBN 0-321-22651-8

STUDENT EDITION OF THOMAS' CALCULUS: EARLY TRANSCENDENTALS, Eleventh Edition
Complete (Chapters 1–16), ISBN 0-321-19800-X
Part One, Single Variable Calculus (Chapters 1–11), ISBN 0-321-22633-X
Part Two, Multivariable Calculus (Chapters 11–16), ISBN 0-321-22651-8
The *Early Transcendentals* version of *Thomas' Calculus* introduces and integrates transcendental functions (such as inverse trigonometric, exponential, and logarithmic functions) into the exposition, examples, and exercises of the early chapters alongside the algebraic functions. Part Two for *Thomas' Calculus: Early Transcendentals* is the same text as Part Two for *Thomas' Calculus.*

Print Supplements

INSTRUCTOR'S SOLUTIONS MANUAL
Part One (Chapters 1–11), ISBN 0-321-22653-4
Part Two (Chapters 11–16), ISBN 0-321-22650-X
The *Instructor's Solutions Manual* by William Ardis, Joseph Borzellino, Linda Buchanan, Alexis T. Mogill, and Patricia Nelson contains complete worked-out solutions to all of the exercises in the text.

ANSWER BOOK
ISBN 0-321-22649-6
The *Answer Book* by William Ardis, Joseph Borzellino, Linda Buchanan, Alexis T. Mogill, and Patricia Nelson contains short answers to most of the exercises in the text.

STUDENT OUTLINES
Part One (Chapters 1–11), ISBN 0-321-22640-2
Part Two (Chapters 11–16), ISBN 0-321-22641-0
Organized to correspond to the text, the *Student Outlines* by Joseph Borzellino and Patricia Nelson reinforces important concepts and provides an outline of the important topics, theorems, and definitions, as well as study tips and additional practice problems.

STUDENT'S SOLUTIONS MANUAL
Part One (Chapters 1–11), ISBN 0-321-22646-1
Part Two (Chapters 11–16), ISBN 0-321-22647-X
The *Student's Solutions Manual* by William Ardis, Joseph Borzellino, Linda Buchanan, Alexis T. Mogill, and Patricia Nelson is designed for the student and contains carefully worked-out solutions to all the odd-numbered exercises in the text.

JUST-IN-TIME ALGEBRA AND TRIGONOMETRY FOR CALCULUS, Third Edition
ISBN 0-321-26943-8
Sharp algebra and trigonometry skills are critical to mastering calculus, and *Just-in-Time Algebra and Trigonometry for Calculus,* Third Edition, by Guntram Mueller and Ronald I. Brent is designed to bolster these skills while students study calculus. As students make their way through calculus, this text is with them every step of the way, showing them the necessary algebra or trigonometry topics and pointing out potential problem spots. The easy-to-use contents has algebra and trigonometry topics arranged in the order in which students will need them as they study calculus.

ADDISON-WESLEY'S CALCULUS REVIEW CARD
The Calculus Review Card is a resource for students containing important formulas, functions, definitions, and theorems that correspond precisely to *Thomas' Calculus.* This card

can work as a reference for completing homework assignments or as an aid in studying and is available bundled with a new text. Contact your Addison-Wesley sales representative for more information.

Media and Online Supplements

TECHNOLOGY RESOURCE MANUALS

Maple Manual by Donald Hartig, California Polytechnic State University
Mathematica Manual by Marie Vanisko, California State University Stanislaus, and Lyle Cochran, Whitworth College
TI-Graphing Calculator Manual by Luz DeAlba, Drake University
These manuals cover *Maple* 9, *Mathematica* 5, and the TI-83 Plus/TI-84 Plus, TI-85/TI-86, and TI-89/TI-92 Plus, respectively. Each manual provides detailed guidance for integrating a specific software package or graphing calculator throughout the course, including syntax and commands. These manuals are available to qualified instructors through **http://suppscentral.aw.com.**

MYMATHLAB ™

MyMathLab is a series of text-specific, easily customizable online courses for Addison-Wesley textbooks in mathematics and statistics. MyMathLab is powered by CourseCompass™—Pearson Education's online teaching and learning environment—and by MathXL—Addison-Wesley's online homework, tutorial, and assessment system. MyMathLab gives you the tools you need to deliver all or a portion of your course online, whether your students are in a lab setting or working from home. MyMathLab provides a rich and flexible set of course materials, featuring free-response exercises that are algorithmically generated for unlimited practice and mastery. Students can also use online tools, such as video lectures, animations, multimedia textbook, and Maple/Mathematica projects, to independently improve their understanding and performance. Instructors can use MyMathLab's homework and test managers to select and assign online exercises correlated directly to the textbook, and they can import TestGen tests into MyMathLab for added flexibility. MyMathLab's online gradebook—designed specifically for mathematics and statistics—automatically tracks students' homework and test results and gives the instructor control over how to calculate final grades. MyMathLab is available to qualified adopters. For more information, visit our Web site at **www.mymathlab.com** or contact your Addison-Wesley sales representative for a product demonstration.

MATHXL®

MathXL is a powerful online homework, tutorial, and assessment system that accompanies your Addison-Wesley textbook in mathematics or statistics. With MathXL, instructors can create, edit, and assign online homework and tests using algorithmically generated exercises correlated at the objective level to the textbook. All student work is tracked in MathXL's online gradebook. Students can take chapter tests in MathXL and receive personalized study plans based on their test results. The study plan diagnoses weaknesses and links students directly to tutorial exercises for the objectives they need to study and retest. Students can also access supplemental animations and video clips directly from selected exercises. MathXL is available to qualified adopters. For more information, visit our Web site at **www.mathxl.com** or contact your Addison-Wesley sales representative for a product demonstration.

TESTGEN WITH QUIZMASTER

TestGen enables instructors to build, edit, print, and administer tests using a computerized bank of questions developed to cover all the objectives of the text. TestGen is algorithmically based, allowing instructors to create multiple but equivalent versions of the same question or test with the click of a button. Instructors can also modify test bank questions or add new questions by using the built-in question editor, which allows users to create graphs, import graphics, and insert math notation, variable numbers, or text. Tests can be printed or administered online via the Internet or another network. TestGen comes packaged with QuizMaster, which allows students to take tests on a local area network. The software is available on a dual-platform Windows/Macintosh CD-ROM.

DIGITAL VIDEO TUTOR

The Digital Video Tutor features an engaging team of mathematics instructors who present comprehensive coverage of topics in the text. The lecturers' presentations include examples and exercises from the text and support an approach that emphasizes visualization and problem solving. The video lectures are available on CD-ROM, making it easy and convenient for students to watch the videos from a computer at home or on campus. The complete digitized video set, affordable and portable for students, is ideal for distance learning or supplemental instruction.

WEB SITE   www.aw-bc.com/thomas

The *Thomas' Calculus* Web site provides the expanded historical biographies and essays referenced in the text. Also available is a collection of *Maple* and *Mathematica* modules that can be used as projects by individual students or groups of students.

ADDISON-WESLEY MATH TUTOR CENTER

The Addison-Wesley Math Tutor Center is staffed by qualified mathematics and statistics instructors who provide students with tutoring on examples and odd-numbered exercises from the textbook. Tutoring is available via toll-free telephone, toll-free fax, e-mail, and the Internet. Interactive, Web-based technology allows tutors and students to view and work through problems together in real time over the Internet. The Addison-Wesley Math Tutor Center is available to qualified adopters. For more information, please visit our Web site at **www.aw-bc.com/tutorcenter** or call us at 1-888-777-0463.

Acknowledgments

We would like to express our thanks to the people who made many valuable contributions to this edition as it developed through its various stages:

Development Editors
  Elka Block
  David Chelton
  Frank Purcell

Accuracy Checkers
  William Ardis
  Karl Kattchee
  Douglas B. Meade
  Robert Pierce
  Frank Purcell
  Marie Vanisko
  Thomas Wegleitner

Super Reviewers
  Harry Allen, *Ohio State University*
  Rebecca Goldin, *George Mason University*
  Christopher Heil, *Georgia Institute of Technology*
  Dominic Naughton, *Purdue University*
  Maria Terrell, *Cornell University*
  Clifford Weil, *Michigan State University*

Reviewers
  Robert Anderson, *University of Wisconsin–Milwaukee*
  Charles Ashley, *Villanova University*
  David Bachman, *California Polytechnic State University*
  Elizabeth Bator, *University of North Texas*

William Bogley, *Oregon State University*

Kaddour Boukaabar, *California University of Pennsylvania*

Deborah Brandon, *Carnegie Mellon University*

Mark Bridger, *Northeastern University*

Sean Cleary, *The City College of New York*

Edward Crotty, *University of Pennsylvania*

Mark Davidson, *Louisiana State University*

Richard Davitt, *University of Louisville*

Elias Deeba, *University of Houston, Downtown Campus*

Anne Dougherty, *University of Colorado*

Rafael Espericueta, *Bakersfield College*

Klaus Fischer, *George Mason University*

William Fitzgibbon, *University of Houston*

Carol Flakus, *Lower Columbia College*

Tim Flood, *Pittsburg State University*

Robert Gardner, *East Tennessee State University*

John Gilbert, *The University of Texas at Austin*

Mark Hanish, *Calvin College*

Zahid Hasan, *California State University, San Bernardino*

Jo W. Heath, *Auburn University*

Ken Holladay, *University of New Orleans*

Hugh Howards, *Wake Forest University*

Dwanye Jennings, *Union University*

Matthias Kawaski, *Arizona State University*

Bill Kincaid, *Wilmington College*

Mark M. Maxwell, *Robert Morris University*

Jack Mealy, *Austin College*

Richard Mercer, *Wright State University*

Victor Nestor, *Pennsylvania State University*

Michael O'Leary, *Towson University*

Bogdan Oporowski, *Louisiana State University*

Troy Riggs, *Union University*

Ferinand Rivera, *San Jose State University*

Mohammed Saleem, *San Jose State University*

Tatiana Shubin, *San Jose State University*

Alex Smith, *University of Wisconsin-Eau Claire*

Donald Solomon, *University of Wisconsin–Milwaukee*

Chia Chi Tung, *Minnesota State University*

William L. VanAlstine, *Aiken Technology College*

Bobby Winters, *Pittsburg State University*

Dennis Wortman, *University of Massachusetts at Boston*

Survey Participants

Omar Adawi, *Parkland College*

Siham Alfred, *Raritan Valley Community College*

Donna J. Bailey, *Truman State University*

Rajesh K. Barnwal, *Middle Tennessee State University*

Robert C. Brigham, *University of Central Florida* (retired)

Thomas A. Carnevale, *Valdosta State University*

Lenny Chastkofsky, *The University of Georgia*

Richard Dalrymple, *Minnesota West Community & Technical College*

Lloyd Davis, *College of San Mateo*

Will-Matthis Dunn III, *Montgomery College*

George F. Feissner, *SUNY College at Cortland*

Bruno Harris, *Brown University*

Celeste Hernandez, *Richland College*

Wei-Min Huang, *Lehigh University*

Herbert E. Kasube, *Bradley University*

Frederick W. Keene, *Pasadena City College*

Michael Kent, *Borough of Manhattan Community College*

Robert Levine, *Community College of Allegheny County, Boyce Campus*

John Martin, *Santa Rosa Junior College*

Michael Scott McClendon, *University of Central Oklahoma*

Ching-Tsuan Pan, *Northern Illinois University*

Emma Previato, *Boston University*

S.S. Ravindran, *University of Alabama*

Dan Rothe, *Alpena Community College*

John T. Saccoman, *Seton Hall University*

Mansour Samimi, *Winston-Salem State University*

Ned W. Schillow, *Lehigh Carbon Community College*

W.R. Schrank, *Angelina College*

Mark R. Woodard, *Furman University*

THOMAS'
CALCULUS

ELEVENTH EDITION

PART TWO

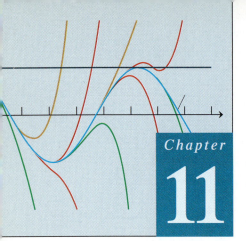

INFINITE SEQUENCES AND SERIES

OVERVIEW While everyone knows how to add together two numbers, or even several, how to add together infinitely many numbers is not so clear. In this chapter we study such questions, the subject of the theory of infinite series. Infinite series sometimes have a finite sum, as in

$$\frac{1}{2} + \frac{1}{4} + \frac{1}{8} + \frac{1}{16} + \cdots = 1.$$

This sum is represented geometrically by the areas of the repeatedly halved unit square shown here. The areas of the small rectangles add together to give the area of the unit square, which they fill. Adding together more and more terms gets us closer and closer to the total.

Other infinite series do not have a finite sum, as with

$$1 + 2 + 3 + 4 + 5 + \cdots.$$

The sum of the first few terms gets larger and larger as we add more and more terms. Taking enough terms makes these sums larger than any prechosen constant.

With some infinite series, such as the harmonic series

$$1 + \frac{1}{2} + \frac{1}{3} + \frac{1}{4} + \frac{1}{5} + \frac{1}{6} + \cdots$$

it is not obvious whether a finite sum exists. It is unclear whether adding more and more terms gets us closer to some sum, or gives sums that grow without bound.

As we develop the theory of infinite sequences and series, an important application gives a method of representing a differentiable function $f(x)$ as an infinite sum of powers of x. With this method we can extend our knowledge of how to evaluate, differentiate, and integrate polynomials to a class of functions much more general than polynomials. We also investigate a method of representing a function as an infinite sum of sine and cosine functions. This method will yield a powerful tool to study functions.

11.1   Sequences

HISTORICAL ESSAY

Sequences and Series

A sequence is a list of numbers

$$a_1, a_2, a_3, \ldots, a_n, \ldots$$

in a given order. Each of a_1, a_2, a_3 and so on represents a number. These are the **terms** of the sequence. For example the sequence

$$2, 4, 6, 8, 10, 12, \ldots, 2n, \ldots$$

has first term $a_1 = 2$, second term $a_2 = 4$ and nth term $a_n = 2n$. The integer n is called the **index** of a_n, and indicates where a_n occurs in the list. We can think of the sequence

$$a_1, a_2, a_3, \ldots, a_n, \ldots$$

as a function that sends 1 to a_1, 2 to a_2, 3 to a_3, and in general sends the positive integer n to the nth term a_n. This leads to the formal definition of a sequence.

DEFINITION   Infinite Sequence

An **infinite sequence** of numbers is a function whose domain is the set of positive integers.

The function associated to the sequence

$$2, 4, 6, 8, 10, 12, \ldots, 2n, \ldots$$

sends 1 to $a_1 = 2$, 2 to $a_2 = 4$, and so on. The general behavior of this sequence is described by the formula

$$a_n = 2n.$$

We can equally well make the domain the integers larger than a given number n_0, and we allow sequences of this type also.

The sequence

$$12, 14, 16, 18, 20, 22 \ldots$$

is described by the formula $a_n = 10 + 2n$. It can also be described by the simpler formula $b_n = 2n$, where the index n starts at 6 and increases. To allow such simpler formulas, we let the first index of the sequence be any integer. In the sequence above, $\{a_n\}$ starts with a_1 while $\{b_n\}$ starts with b_6. Order is important. The sequence $1, 2, 3, 4 \ldots$ is not the same as the sequence $2, 1, 3, 4 \ldots$.

Sequences can be described by writing rules that specify their terms, such as

$$a_n = \sqrt{n},$$
$$b_n = (-1)^{n+1} \frac{1}{n},$$
$$c_n = \frac{n-1}{n},$$
$$d_n = (-1)^{n+1}$$

or by listing terms,

$$\{a_n\} = \left\{ \sqrt{1}, \sqrt{2}, \sqrt{3}, \ldots, \sqrt{n}, \ldots \right\}$$

$$\{b_n\} = \left\{ 1, -\frac{1}{2}, \frac{1}{3}, -\frac{1}{4}, \ldots, (-1)^{n+1}\frac{1}{n}, \ldots \right\}$$

$$\{c_n\} = \left\{ 0, \frac{1}{2}, \frac{2}{3}, \frac{3}{4}, \frac{4}{5}, \ldots, \frac{n-1}{n}, \ldots \right\}$$

$$\{d_n\} = \left\{ 1, -1, 1, -1, 1, -1, \ldots, (-1)^{n+1}, \ldots \right\}.$$

We also sometimes write

$$\{a_n\} = \left\{ \sqrt{n} \right\}_{n=1}^{\infty}.$$

Figure 11.1 shows two ways to represent sequences graphically. The first marks the first few points from $a_1, a_2, a_3, \ldots, a_n, \ldots$ on the real axis. The second method shows the graph of the function defining the sequence. The function is defined only on integer inputs, and the graph consists of some points in the xy-plane, located at $(1, a_1)$, $(2, a_2), \ldots, (n, a_n), \ldots$.

FIGURE 11.1 Sequences can be represented as points on the real line or as points in the plane where the horizontal axis n is the index number of the term and the vertical axis a_n is its value.

Convergence and Divergence

Sometimes the numbers in a sequence approach a single value as the index n increases. This happens in the sequence

$$\left\{ 1, \frac{1}{2}, \frac{1}{3}, \frac{1}{4}, \ldots, \frac{1}{n}, \ldots \right\}$$

whose terms approach 0 as n gets large, and in the sequence

$$\left\{ 0, \frac{1}{2}, \frac{2}{3}, \frac{3}{4}, \frac{4}{5}, \ldots, 1 - \frac{1}{n}, \ldots \right\}$$

whose terms approach 1. On the other hand, sequences like

$$\left\{\sqrt{1}, \sqrt{2}, \sqrt{3}, \ldots, \sqrt{n}, \ldots\right\}$$

have terms that get larger than any number as n increases, and sequences like

$$\{1, -1, 1, -1, 1, -1, \ldots, (-1)^{n+1}, \ldots\}$$

bounce back and forth between 1 and -1, never converging to a single value. The following definition captures the meaning of having a sequence converge to a limiting value. It says that if we go far enough out in the sequence, by taking the index n to be larger then some value N, the difference between a_n and the limit of the sequence becomes less than any preselected number $\epsilon > 0$.

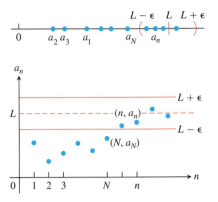

FIGURE 11.2 $a_n \to L$ if $y = L$ is a horizontal asymptote of the sequence of points $\{(n, a_n)\}$. In this figure, all the a_n's after a_N lie within ϵ of L.

DEFINITIONS Converges, Diverges, Limit

The sequence $\{a_n\}$ **converges** to the number L if to every positive number ϵ there corresponds an integer N such that for all n,

$$n > N \quad \Rightarrow \quad |a_n - L| < \epsilon.$$

If no such number L exists, we say that $\{a_n\}$ **diverges**.

If $\{a_n\}$ converges to L, we write $\lim_{n\to\infty} a_n = L$, or simply $a_n \to L$, and call L the **limit** of the sequence (Figure 11.2).

The definition is very similar to the definition of the limit of a function $f(x)$ as x tends to ∞ ($\lim_{x\to\infty} f(x)$ in Section 2.4). We will exploit this connection to calculate limits of sequences.

EXAMPLE 1 Applying the Definition

Show that

(a) $\lim\limits_{n\to\infty} \dfrac{1}{n} = 0$ (b) $\lim\limits_{n\to\infty} k = k$ (any constant k)

Solution

(a) Let $\epsilon > 0$ be given. We must show that there exists an integer N such that for all n,

$$n > N \quad \Rightarrow \quad \left|\frac{1}{n} - 0\right| < \epsilon.$$

This implication will hold if $(1/n) < \epsilon$ or $n > 1/\epsilon$. If N is any integer greater than $1/\epsilon$, the implication will hold for all $n > N$. This proves that $\lim_{n\to\infty}(1/n) = 0$.

(b) Let $\epsilon > 0$ be given. We must show that there exists an integer N such that for all n,

$$n > N \quad \Rightarrow \quad |k - k| < \epsilon.$$

Since $k - k = 0$, we can use any positive integer for N and the implication will hold. This proves that $\lim_{n\to\infty} k = k$ for any constant k. ∎

EXAMPLE 2   A Divergent Sequence

Show that the sequence $\{1, -1, 1, -1, 1, -1, \ldots, (-1)^{n+1}, \ldots\}$ diverges.

Solution   Suppose the sequence converges to some number L. By choosing $\epsilon = 1/2$ in the definition of the limit, all terms a_n of the sequence with index n larger than some N must lie within $\epsilon = 1/2$ of L. Since the number 1 appears repeatedly as every other term of the sequence, we must have that the number 1 lies within the distance $\epsilon = 1/2$ of L. It follows that $|L - 1| < 1/2$, or equivalently, $1/2 < L < 3/2$. Likewise, the number -1 appears repeatedly in the sequence with arbitrarily high index. So we must also have that $|L - (-1)| < 1/2$, or equivalently, $-3/2 < L < -1/2$. But the number L cannot lie in both of the intervals $(1/2, 3/2)$ and $(-3/2, -1/2)$ because they have no overlap. Therefore, no such limit L exists and so the sequence diverges.

Note that the same argument works for any positive number ϵ smaller than 1, not just 1/2. ∎

The sequence $\{\sqrt{n}\}$ also diverges, but for a different reason. As n increases, its terms become larger than any fixed number. We describe the behavior of this sequence by writing

$$\lim_{n \to \infty} \sqrt{n} = \infty.$$

In writing infinity as the limit of a sequence, we are not saying that the differences between the terms a_n and ∞ become small as n increases. Nor are we asserting that there is some number infinity that the sequence approaches. We are merely using a notation that captures the idea that a_n eventually gets and stays larger than any fixed number as n gets large.

DEFINITION   **Diverges to Infinity**

The sequence $\{a_n\}$ **diverges to infinity** if for every number M there is an integer N such that for all n larger than N, $a_n > M$. If this condition holds we write

$$\lim_{n \to \infty} a_n = \infty \quad \text{or} \quad a_n \to \infty.$$

Similarly if for every number m there is an integer N such that for all $n > N$ we have $a_n < m$, then we say $\{a_n\}$ **diverges to negative infinity** and write

$$\lim_{n \to \infty} a_n = -\infty \quad \text{or} \quad a_n \to -\infty.$$

A sequence may diverge without diverging to infinity or negative infinity. We saw this in Example 2, and the sequences $\{1, -2, 3, -4, 5, -6, 7, -8, \ldots\}$ and $\{1, 0, 2, 0, 3, 0, \ldots\}$ are also examples of such divergence.

Calculating Limits of Sequences

If we always had to use the formal definition of the limit of a sequence, calculating with ϵ's and N's, then computing limits of sequences would be a formidable task. Fortunately we can derive a few basic examples, and then use these to quickly analyze the limits of many more sequences. We will need to understand how to combine and compare sequences. Since sequences are functions with domain restricted to the positive integers, it is not too surprising that the theorems on limits of functions given in Chapter 2 have versions for sequences.

THEOREM 1

Let $\{a_n\}$ and $\{b_n\}$ be sequences of real numbers and let A and B be real numbers. The following rules hold if $\lim_{n\to\infty} a_n = A$ and $\lim_{n\to\infty} b_n = B$.

1. *Sum Rule:* $\lim_{n\to\infty}(a_n + b_n) = A + B$

2. *Difference Rule:* $\lim_{n\to\infty}(a_n - b_n) = A - B$

3. *Product Rule:* $\lim_{n\to\infty}(a_n \cdot b_n) = A \cdot B$

4. *Constant Multiple Rule:* $\lim_{n\to\infty}(k \cdot b_n) = k \cdot B$ (Any number k)

5. *Quotient Rule:* $\lim_{n\to\infty} \dfrac{a_n}{b_n} = \dfrac{A}{B}$ if $B \neq 0$

The proof is similar to that of Theorem 1 of Section 2.2, and is omitted.

EXAMPLE 3   Applying Theorem 1

By combining Theorem 1 with the limits of Example 1, we have:

(a) $\lim_{n\to\infty}\left(-\dfrac{1}{n}\right) = -1 \cdot \lim_{n\to\infty}\dfrac{1}{n} = -1 \cdot 0 = 0$ Constant Multiple Rule and Example 1a

(b) $\lim_{n\to\infty}\left(\dfrac{n-1}{n}\right) = \lim_{n\to\infty}\left(1 - \dfrac{1}{n}\right) = \lim_{n\to\infty} 1 - \lim_{n\to\infty}\dfrac{1}{n} = 1 - 0 = 1$ Difference Rule and Example 1a

(c) $\lim_{n\to\infty}\dfrac{5}{n^2} = 5 \cdot \lim_{n\to\infty}\dfrac{1}{n} \cdot \lim_{n\to\infty}\dfrac{1}{n} = 5 \cdot 0 \cdot 0 = 0$ Product Rule

(d) $\lim_{n\to\infty}\dfrac{4 - 7n^6}{n^6 + 3} = \lim_{n\to\infty}\dfrac{(4/n^6) - 7}{1 + (3/n^6)} = \dfrac{0 - 7}{1 + 0} = -7.$ Sum and Quotient Rules   ∎

Be cautious in applying Theorem 1. It does not say, for example, that each of the sequences $\{a_n\}$ and $\{b_n\}$ have limits if their sum $\{a_n + b_n\}$ has a limit. For instance, $\{a_n\} = \{1, 2, 3, \ldots\}$ and $\{b_n\} = \{-1, -2, -3, \ldots\}$ both diverge, but their sum $\{a_n + b_n\} = \{0, 0, 0, \ldots\}$ clearly converges to 0.

One consequence of Theorem 1 is that every nonzero multiple of a divergent sequence $\{a_n\}$ diverges. For suppose, to the contrary, that $\{ca_n\}$ converges for some number $c \neq 0$. Then, by taking $k = 1/c$ in the Constant Multiple Rule in Theorem 1, we see that the sequence

$$\left\{\dfrac{1}{c} \cdot ca_n\right\} = \{a_n\}$$

converges. Thus, $\{ca_n\}$ cannot converge unless $\{a_n\}$ also converges. If $\{a_n\}$ does not converge, then $\{ca_n\}$ does not converge.

The next theorem is the sequence version of the Sandwich Theorem in Section 2.2. You are asked to prove the theorem in Exercise 95.

THEOREM 2   The Sandwich Theorem for Sequences

Let $\{a_n\}$, $\{b_n\}$, and $\{c_n\}$ be sequences of real numbers. If $a_n \leq b_n \leq c_n$ holds for all n beyond some index N, and if $\lim_{n\to\infty} a_n = \lim_{n\to\infty} c_n = L$, then $\lim_{n\to\infty} b_n = L$ also.

An immediate consequence of Theorem 2 is that, if $|b_n| \le c_n$ and $c_n \to 0$, then $b_n \to 0$ because $-c_n \le b_n \le c_n$. We use this fact in the next example.

EXAMPLE 4   Applying the Sandwich Theorem

Since $1/n \to 0$, we know that

(a) $\dfrac{\cos n}{n} \to 0$ because $-\dfrac{1}{n} \le \dfrac{\cos n}{n} \le \dfrac{1}{n}$;

(b) $\dfrac{1}{2^n} \to 0$ because $0 \le \dfrac{1}{2^n} \le \dfrac{1}{n}$;

(c) $(-1)^n \dfrac{1}{n} \to 0$ because $-\dfrac{1}{n} \le (-1)^n \dfrac{1}{n} \le \dfrac{1}{n}$.   ∎

The application of Theorems 1 and 2 is broadened by a theorem stating that applying a continuous function to a convergent sequence produces a convergent sequence. We state the theorem without proof (Exercise 96).

THEOREM 3   The Continuous Function Theorem for Sequences

Let $\{a_n\}$ be a sequence of real numbers. If $a_n \to L$ and if f is a function that is continuous at L and defined at all a_n, then $f(a_n) \to f(L)$.

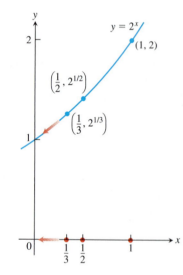

FIGURE 11.3 As $n \to \infty$, $1/n \to 0$ and $2^{1/n} \to 2^0$ (Example 6).

EXAMPLE 5   Applying Theorem 3

Show that $\sqrt{(n+1)/n} \to 1$.

Solution   We know that $(n+1)/n \to 1$. Taking $f(x) = \sqrt{x}$ and $L = 1$ in Theorem 3 gives $\sqrt{(n+1)/n} \to \sqrt{1} = 1$.   ∎

EXAMPLE 6   The Sequence $\{2^{1/n}\}$

The sequence $\{1/n\}$ converges to 0. By taking $a_n = 1/n$, $f(x) = 2^x$, and $L = 0$ in Theorem 3, we see that $2^{1/n} = f(1/n) \to f(L) = 2^0 = 1$. The sequence $\{2^{1/n}\}$ converges to 1 (Figure 11.3).   ∎

Using l'Hôpital's Rule

The next theorem enables us to use l'Hôpital's Rule to find the limits of some sequences. It formalizes the connection between $\lim_{n \to \infty} a_n$ and $\lim_{x \to \infty} f(x)$.

THEOREM 4

Suppose that $f(x)$ is a function defined for all $x \ge n_0$ and that $\{a_n\}$ is a sequence of real numbers such that $a_n = f(n)$ for $n \ge n_0$. Then

$$\lim_{x \to \infty} f(x) = L \quad \Longrightarrow \quad \lim_{n \to \infty} a_n = L.$$

Proof   Suppose that $\lim_{x \to \infty} f(x) = L$. Then for each positive number ϵ there is a number M such that for all x,

$$x > M \quad \Longrightarrow \quad |f(x) - L| < \epsilon.$$

Let N be an integer greater than M and greater than or equal to n_0. Then

$$n > N \quad \Rightarrow \quad a_n = f(n) \quad \text{and} \quad |a_n - L| = |f(n) - L| < \epsilon. \qquad \blacksquare$$

EXAMPLE 7 Applying L'Hôpital's Rule

Show that

$$\lim_{n \to \infty} \frac{\ln n}{n} = 0.$$

Solution The function $(\ln x)/x$ is defined for all $x \geq 1$ and agrees with the given sequence at positive integers. Therefore, by Theorem 5, $\lim_{n \to \infty}(\ln n)/n$ will equal $\lim_{x \to \infty}(\ln x)/x$ if the latter exists. A single application of l'Hôpital's Rule shows that

$$\lim_{x \to \infty} \frac{\ln x}{x} = \lim_{x \to \infty} \frac{1/x}{1} = \frac{0}{1} = 0.$$

We conclude that $\lim_{n \to \infty}(\ln n)/n = 0$. $\qquad \blacksquare$

When we use l'Hôpital's Rule to find the limit of a sequence, we often treat n as a continuous real variable and differentiate directly with respect to n. This saves us from having to rewrite the formula for a_n as we did in Example 7.

EXAMPLE 8 Applying L'Hôpital's Rule

Find

$$\lim_{n \to \infty} \frac{2^n}{5n}.$$

Solution By l'Hôpital's Rule (differentiating with respect to n),

$$\lim_{n \to \infty} \frac{2^n}{5n} = \lim_{n \to \infty} \frac{2^n \cdot \ln 2}{5}$$
$$= \infty. \qquad \blacksquare$$

EXAMPLE 9 Applying L'Hôpital's Rule to Determine Convergence

Does the sequence whose nth term is

$$a_n = \left(\frac{n + 1}{n - 1}\right)^n$$

converge? If so, find $\lim_{n \to \infty} a_n$.

Solution The limit leads to the indeterminate form 1^∞. We can apply l'Hôpital's Rule if we first change the form to $\infty \cdot 0$ by taking the natural logarithm of a_n:

$$\ln a_n = \ln \left(\frac{n + 1}{n - 1}\right)^n$$
$$= n \ln \left(\frac{n + 1}{n - 1}\right).$$

Then,

$$\lim_{n \to \infty} \ln a_n = \lim_{n \to \infty} n \ln \left(\frac{n + 1}{n - 1} \right) \qquad \infty \cdot 0$$

$$= \lim_{n \to \infty} \frac{\ln \left(\dfrac{n + 1}{n - 1} \right)}{1/n} \qquad \frac{0}{0}$$

$$= \lim_{n \to \infty} \frac{-2/(n^2 - 1)}{-1/n^2} \qquad \text{l'Hôpital's Rule}$$

$$= \lim_{n \to \infty} \frac{2n^2}{n^2 - 1} = 2.$$

Since $\ln a_n \to 2$ and $f(x) = e^x$ is continuous, Theorem 4 tells us that

$$a_n = e^{\ln a_n} \to e^2.$$

The sequence $\{a_n\}$ converges to e^2. ■

Commonly Occurring Limits

The next theorem gives some limits that arise frequently.

THEOREM 5

The following six sequences converge to the limits listed below:

1. $\displaystyle \lim_{n \to \infty} \frac{\ln n}{n} = 0$

2. $\displaystyle \lim_{n \to \infty} \sqrt[n]{n} = 1$

3. $\displaystyle \lim_{n \to \infty} x^{1/n} = 1 \qquad (x > 0)$

4. $\displaystyle \lim_{n \to \infty} x^n = 0 \qquad (|x| < 1)$

5. $\displaystyle \lim_{n \to \infty} \left(1 + \frac{x}{n} \right)^n = e^x \qquad (\text{any } x)$

6. $\displaystyle \lim_{n \to \infty} \frac{x^n}{n!} = 0 \qquad (\text{any } x)$

In Formulas (3) through (6), x remains fixed as $n \to \infty$.

Proof   The first limit was computed in Example 7. The next two can be proved by taking logarithms and applying Theorem 4 (Exercises 93 and 94). The remaining proofs are given in Appendix 3.   ■

EXAMPLE 10   Applying Theorem 5

(a) $\dfrac{\ln (n^2)}{n} = \dfrac{2 \ln n}{n} \to 2 \cdot 0 = 0$ Formula 1

(b) $\sqrt[n]{n^2} = n^{2/n} = (n^{1/n})^2 \to (1)^2 = 1$ Formula 2

(c) $\sqrt[n]{3n} = 3^{1/n}(n^{1/n}) \to 1 \cdot 1 = 1$ Formula 3 with $x = 3$ and Formula 2

Factorial Notation

The notation $n!$ ("n factorial") means the product $1 \cdot 2 \cdot 3 \cdots n$ of the integers from 1 to n. Notice that $(n + 1)! = (n + 1) \cdot n!$. Thus, $4! = 1 \cdot 2 \cdot 3 \cdot 4 = 24$ and $5! = 1 \cdot 2 \cdot 3 \cdot 4 \cdot 5 = 5 \cdot 4! = 120$. We define $0!$ to be 1. Factorials grow even faster than exponentials, as the table suggests.

n	e^n (rounded)	$n!$
1	3	1
5	148	120
10	22,026	3,628,800
20	4.9×10^8	2.4×10^{18}

(d) $\left(-\dfrac{1}{2}\right)^n \to 0$ 　　　　　　　Formula 4 with $x = -\dfrac{1}{2}$

(e) $\left(\dfrac{n-2}{n}\right)^n = \left(1 + \dfrac{-2}{n}\right)^n \to e^{-2}$ 　　Formula 5 with $x = -2$

(f) $\dfrac{100^n}{n!} \to 0$ 　　　　　　　　　Formula 6 with $x = 100$ 　■

Recursive Definitions

So far, we have calculated each a_n directly from the value of n. But sequences are often defined **recursively** by giving

1. The value(s) of the initial term or terms, and

2. A rule, called a **recursion formula**, for calculating any later term from terms that precede it.

EXAMPLE 11　Sequences Constructed Recursively

(a) The statements $a_1 = 1$ and $a_n = a_{n-1} + 1$ define the sequence $1, 2, 3, \ldots, n, \ldots$ of positive integers. With $a_1 = 1$, we have $a_2 = a_1 + 1 = 2, a_3 = a_2 + 1 = 3$, and so on.

(b) The statements $a_1 = 1$ and $a_n = n \cdot a_{n-1}$ define the sequence $1, 2, 6, 24, \ldots, n!, \ldots$ of factorials. With $a_1 = 1$, we have $a_2 = 2 \cdot a_1 = 2, a_3 = 3 \cdot a_2 = 6, a_4 = 4 \cdot a_3 = 24$, and so on.

(c) The statements $a_1 = 1, a_2 = 1$, and $a_{n+1} = a_n + a_{n-1}$ define the sequence $1, 1, 2, 3, 5, \ldots$ of **Fibonacci numbers**. With $a_1 = 1$ and $a_2 = 1$, we have $a_3 = 1 + 1 = 2, a_4 = 2 + 1 = 3, a_5 = 3 + 2 = 5$, and so on.

(d) As we can see by applying Newton's method, the statements $x_0 = 1$ and $x_{n+1} = x_n - [(\sin x_n - x_n^2)/(\cos x_n - 2x_n)]$ define a sequence that converges to a solution of the equation $\sin x - x^2 = 0$. 　■

Bounded Nondecreasing Sequences

The terms of a general sequence can bounce around, sometimes getting larger, sometimes smaller. An important special kind of sequence is one for which each term is at least as large as its predecessor.

> **DEFINITION　Nondecreasing Sequence**
>
> A sequence $\{a_n\}$ with the property that $a_n \le a_{n+1}$ for all n is called a **nondecreasing sequence**.

EXAMPLE 12　Nondecreasing Sequences

(a) The sequence $1, 2, 3, \ldots, n, \ldots$ of natural numbers

(b) The sequence $\dfrac{1}{2}, \dfrac{2}{3}, \dfrac{3}{4}, \ldots, \dfrac{n}{n+1}, \ldots$

(c) The constant sequence $\{3\}$ 　■

There are two kinds of nondecreasing sequences—those whose terms increase beyond any finite bound and those whose terms do not.

DEFINITIONS   Bounded, Upper Bound, Least Upper Bound

A sequence $\{a_n\}$ is **bounded from above** if there exists a number M such that $a_n \leq M$ for all n. The number M is an **upper bound** for $\{a_n\}$. If M is an upper bound for $\{a_n\}$ but no number less than M is an upper bound for $\{a_n\}$, then M is the **least upper bound** for $\{a_n\}$.

EXAMPLE 13   Applying the Definition for Boundedness

(a) The sequence $1, 2, 3, \ldots, n, \ldots$ has no upper bound.

(b) The sequence $\dfrac{1}{2}, \dfrac{2}{3}, \dfrac{3}{4}, \ldots, \dfrac{n}{n+1}, \ldots$ is bounded above by $M = 1$.

No number less than 1 is an upper bound for the sequence, so 1 is the least upper bound (Exercise 113). ■

A nondecreasing sequence that is bounded from above always has a least upper bound. This is the completeness property of the real numbers, discussed in Appendix 4. We will prove that if L is the least upper bound then the sequence converges to L.

Suppose we plot the points $(1, a_1), (2, a_2), \ldots, (n, a_n), \ldots$ in the xy-plane. If M is an upper bound of the sequence, all these points will lie on or below the line $y = M$ (Figure 11.4). The line $y = L$ is the lowest such line. None of the points (n, a_n) lies above $y = L$, but some do lie above any lower line $y = L - \epsilon$, if ϵ is a positive number. The sequence converges to L because

(a) $a_n \leq L$ for *all* values of n and

(b) given any $\epsilon > 0$, there exists at least one integer N for which $a_N > L - \epsilon$.

The fact that $\{a_n\}$ is nondecreasing tells us further that

$$a_n \geq a_N > L - \epsilon \qquad \text{for all } n \geq N.$$

Thus, *all* the numbers a_n beyond the Nth number lie within ϵ of L. This is precisely the condition for L to be the limit of the sequence $\{a_n\}$.

The facts for nondecreasing sequences are summarized in the following theorem. A similar result holds for nonincreasing sequences (Exercise 107).

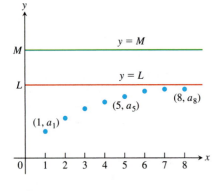

FIGURE 11.4   If the terms of a nondecreasing sequence have an upper bound M, they have a limit $L \leq M$.

THEOREM 6   The Nondecreasing Sequence Theorem

A nondecreasing sequence of real numbers converges if and only if it is bounded from above. If a nondecreasing sequence converges, it converges to its least upper bound.

Theorem 6 implies that a nondecreasing sequence converges when it is bounded from above. It diverges to infinity if it is not bounded from above.

EXERCISES 11.1

Finding Terms of a Sequence

Each of Exercises 1–6 gives a formula for the nth term a_n of a sequence $\{a_n\}$. Find the values of $a_1, a_2, a_3,$ and a_4.

1. $a_n = \dfrac{1 - n}{n^2}$

2. $a_n = \dfrac{1}{n!}$

3. $a_n = \dfrac{(-1)^{n+1}}{2n - 1}$

4. $a_n = 2 + (-1)^n$

5. $a_n = \dfrac{2^n}{2^{n+1}}$

6. $a_n = \dfrac{2^n - 1}{2^n}$

Each of Exercises 7–12 gives the first term or two of a sequence along with a recursion formula for the remaining terms. Write out the first ten terms of the sequence.

7. $a_1 = 1, \quad a_{n+1} = a_n + (1/2^n)$

8. $a_1 = 1, \quad a_{n+1} = a_n/(n + 1)$

9. $a_1 = 2, \quad a_{n+1} = (-1)^{n+1}a_n/2$

10. $a_1 = -2, \quad a_{n+1} = na_n/(n + 1)$

11. $a_1 = a_2 = 1, \quad a_{n+2} = a_{n+1} + a_n$

12. $a_1 = 2, \quad a_2 = -1, \quad a_{n+2} = a_{n+1}/a_n$

Finding a Sequence's Formula

In Exercises 13–22, find a formula for the nth term of the sequence.

13. The sequence $1, -1, 1, -1, 1, \ldots$ 1's with alternating signs

14. The sequence $-1, 1, -1, 1, -1, \ldots$ 1's with alternating signs

15. The sequence $1, -4, 9, -16, 25, \ldots$ Squares of the positive integers; with alternating signs

16. The sequence $1, -\dfrac{1}{4}, \dfrac{1}{9}, -\dfrac{1}{16}, \dfrac{1}{25}, \ldots$ Reciprocals of squares of the positive integers, with alternating signs

17. The sequence $0, 3, 8, 15, 24, \ldots$ Squares of the positive integers diminished by 1

18. The sequence $-3, -2, -1, 0, 1, \ldots$ Integers beginning with -3

19. The sequence $1, 5, 9, 13, 17, \ldots$ Every other odd positive integer

20. The sequence $2, 6, 10, 14, 18, \ldots$ Every other even positive integer

21. The sequence $1, 0, 1, 0, 1, \ldots$ Alternating 1's and 0's

22. The sequence $0, 1, 1, 2, 2, 3, 3, 4, \ldots$ Each positive integer repeated

Finding Limits

Which of the sequences $\{a_n\}$ in Exercises 23–84 converge, and which diverge? Find the limit of each convergent sequence.

23. $a_n = 2 + (0.1)^n$

24. $a_n = \dfrac{n + (-1)^n}{n}$

25. $a_n = \dfrac{1 - 2n}{1 + 2n}$

26. $a_n = \dfrac{2n + 1}{1 - 3\sqrt{n}}$

27. $a_n = \dfrac{1 - 5n^4}{n^4 + 8n^3}$

28. $a_n = \dfrac{n + 3}{n^2 + 5n + 6}$

29. $a_n = \dfrac{n^2 - 2n + 1}{n - 1}$

30. $a_n = \dfrac{1 - n^3}{70 - 4n^2}$

31. $a_n = 1 + (-1)^n$

32. $a_n = (-1)^n \left(1 - \dfrac{1}{n}\right)$

33. $a_n = \left(\dfrac{n + 1}{2n}\right)\left(1 - \dfrac{1}{n}\right)$

34. $a_n = \left(2 - \dfrac{1}{2^n}\right)\left(3 + \dfrac{1}{2^n}\right)$

35. $a_n = \dfrac{(-1)^{n+1}}{2n - 1}$

36. $a_n = \left(-\dfrac{1}{2}\right)^n$

37. $a_n = \sqrt{\dfrac{2n}{n + 1}}$

38. $a_n = \dfrac{1}{(0.9)^n}$

39. $a_n = \sin\left(\dfrac{\pi}{2} + \dfrac{1}{n}\right)$

40. $a_n = n\pi \cos(n\pi)$

41. $a_n = \dfrac{\sin n}{n}$

42. $a_n = \dfrac{\sin^2 n}{2^n}$

43. $a_n = \dfrac{n}{2^n}$

44. $a_n = \dfrac{3^n}{n^3}$

45. $a_n = \dfrac{\ln(n + 1)}{\sqrt{n}}$

46. $a_n = \dfrac{\ln n}{\ln 2n}$

47. $a_n = 8^{1/n}$

48. $a_n = (0.03)^{1/n}$

49. $a_n = \left(1 + \dfrac{7}{n}\right)^n$

50. $a_n = \left(1 - \dfrac{1}{n}\right)^n$

51. $a_n = \sqrt[n]{10n}$

52. $a_n = \sqrt[n]{n^2}$

53. $a_n = \left(\dfrac{3}{n}\right)^{1/n}$

54. $a_n = (n + 4)^{1/(n+4)}$

55. $a_n = \dfrac{\ln n}{n^{1/n}}$

56. $a_n = \ln n - \ln(n + 1)$

57. $a_n = \sqrt[n]{4^n n}$

58. $a_n = \sqrt[n]{3^{2n+1}}$

59. $a_n = \dfrac{n!}{n^n}$ (*Hint:* Compare with $1/n$.)

60. $a_n = \dfrac{(-4)^n}{n!}$

61. $a_n = \dfrac{n!}{10^{6n}}$

62. $a_n = \dfrac{n!}{2^n \cdot 3^n}$

63. $a_n = \left(\dfrac{1}{n}\right)^{1/(\ln n)}$

64. $a_n = \ln\left(1 + \dfrac{1}{n}\right)^n$

65. $a_n = \left(\dfrac{3n+1}{3n-1}\right)^n$

66. $a_n = \left(\dfrac{n}{n+1}\right)^n$

67. $a_n = \left(\dfrac{x^n}{2n+1}\right)^{1/n}, \quad x > 0$

68. $a_n = \left(1 - \dfrac{1}{n^2}\right)^n$

69. $a_n = \dfrac{3^n \cdot 6^n}{2^{-n} \cdot n!}$

70. $a_n = \dfrac{(10/11)^n}{(9/10)^n + (11/12)^n}$

71. $a_n = \tanh n$

72. $a_n = \sinh(\ln n)$

73. $a_n = \dfrac{n^2}{2n-1} \sin\dfrac{1}{n}$

74. $a_n = n\left(1 - \cos\dfrac{1}{n}\right)$

75. $a_n = \tan^{-1} n$

76. $a_n = \dfrac{1}{\sqrt{n}}\tan^{-1} n$

77. $a_n = \left(\dfrac{1}{3}\right)^n + \dfrac{1}{\sqrt{2^n}}$

78. $a_n = \sqrt[n]{n^2 + n}$

79. $a_n = \dfrac{(\ln n)^{200}}{n}$

80. $a_n = \dfrac{(\ln n)^5}{\sqrt{n}}$

81. $a_n = n - \sqrt{n^2 - n}$

82. $a_n = \dfrac{1}{\sqrt{n^2 - 1} - \sqrt{n^2 + n}}$

83. $a_n = \dfrac{1}{n}\displaystyle\int_1^n \dfrac{1}{x}\,dx$

84. $a_n = \displaystyle\int_1^n \dfrac{1}{x^p}\,dx, \quad p > 1$

Theory and Examples

85. The first term of a sequence is $x_1 = 1$. Each succeeding term is the sum of all those that come before it:

$$x_{n+1} = x_1 + x_2 + \cdots + x_n.$$

Write out enough early terms of the sequence to deduce a general formula for x_n that holds for $n \geq 2$.

86. A sequence of rational numbers is described as follows:

$$\frac{1}{1}, \frac{3}{2}, \frac{7}{5}, \frac{17}{12}, \ldots, \frac{a}{b}, \frac{a+2b}{a+b}, \ldots.$$

Here the numerators form one sequence, the denominators form a second sequence, and their ratios form a third sequence. Let x_n and y_n be, respectively, the numerator and the denominator of the nth fraction $r_n = x_n/y_n$.

a. Verify that $x_1^2 - 2y_1^2 = -1, x_2^2 - 2y_2^2 = +1$ and, more generally, that if $a^2 - 2b^2 = -1$ or $+1$, then

$$(a + 2b)^2 - 2(a + b)^2 = +1 \quad \text{or} \quad -1,$$

respectively.

b. The fractions $r_n = x_n/y_n$ approach a limit as n increases. What is that limit? (*Hint:* Use part (a) to show that $r_n^2 - 2 = \pm(1/y_n)^2$ and that y_n is not less than n.)

87. Newton's method The following sequences come from the recursion formula for Newton's method,

$$x_{n+1} = x_n - \frac{f(x_n)}{f'(x_n)}.$$

Do the sequences converge? If so, to what value? In each case, begin by identifying the function f that generates the sequence.

a. $x_0 = 1, \quad x_{n+1} = x_n - \dfrac{x_n^2 - 2}{2x_n} = \dfrac{x_n}{2} + \dfrac{1}{x_n}$

b. $x_0 = 1, \quad x_{n+1} = x_n - \dfrac{\tan x_n - 1}{\sec^2 x_n}$

c. $x_0 = 1, \quad x_{n+1} = x_n - 1$

88. a. Suppose that $f(x)$ is differentiable for all x in $[0, 1]$ and that $f(0) = 0$. Define the sequence $\{a_n\}$ by the rule $a_n = nf(1/n)$. Show that $\lim_{n\to\infty} a_n = f'(0)$.

Use the result in part (a) to find the limits of the following sequences $\{a_n\}$.

b. $a_n = n\tan^{-1}\dfrac{1}{n}$

c. $a_n = n(e^{1/n} - 1)$

d. $a_n = n\ln\left(1 + \dfrac{2}{n}\right)$

89. Pythagorean triples A triple of positive integers a, b, and c is called a **Pythagorean triple** if $a^2 + b^2 = c^2$. Let a be an odd positive integer and let

$$b = \left\lfloor \frac{a^2}{2} \right\rfloor \quad \text{and} \quad c = \left\lceil \frac{a^2}{2} \right\rceil$$

be, respectively, the integer floor and ceiling for $a^2/2$.

a. Show that $a^2 + b^2 = c^2$. (*Hint:* Let $a = 2n + 1$ and express b and c in terms of n.)

b. By direct calculation, or by appealing to the figure here, find

$$\lim_{a \to \infty} \frac{\left\lfloor \dfrac{a^2}{2} \right\rfloor}{\left\lceil \dfrac{a^2}{2} \right\rceil}.$$

90. The nth root of $n!$

a. Show that $\lim_{n \to \infty} (2n\pi)^{1/(2n)} = 1$ and hence, using Stirling's approximation (Chapter 8, Additional Exercise 50a), that

$$\sqrt[n]{n!} \approx \frac{n}{e} \quad \text{for large values of } n.$$

T **b.** Test the approximation in part (a) for $n = 40, 50, 60, \ldots$, as far as your calculator will allow.

91. a. Assuming that $\lim_{n \to \infty} (1/n^c) = 0$ if c is any positive constant, show that

$$\lim_{n \to \infty} \frac{\ln n}{n^c} = 0$$

if c is any positive constant.

b. Prove that $\lim_{n \to \infty} (1/n^c) = 0$ if c is any positive constant. (*Hint:* If $\epsilon = 0.001$ and $c = 0.04$, how large should N be to ensure that $|1/n^c - 0| < \epsilon$ if $n > N$?)

92. The zipper theorem Prove the "zipper theorem" for sequences: If $\{a_n\}$ and $\{b_n\}$ both converge to L, then the sequence

$$a_1, b_1, a_2, b_2, \ldots, a_n, b_n, \ldots$$

converges to L.

93. Prove that $\lim_{n \to \infty} \sqrt[n]{n} = 1$.

94. Prove that $\lim_{n \to \infty} x^{1/n} = 1, (x > 0)$.

95. Prove Theorem 2.   **96.** Prove Theorem 3.

In Exercises 97–100, determine if the sequence is nondecreasing and if it is bounded from above.

97. $a_n = \dfrac{3n+1}{n+1}$ **98.** $a_n = \dfrac{(2n+3)!}{(n+1)!}$

99. $a_n = \dfrac{2^n 3^n}{n!}$ **100.** $a_n = 2 - \dfrac{2}{n} - \dfrac{1}{2^n}$

Which of the sequences in Exercises 101–106 converge, and which diverge? Give reasons for your answers.

101. $a_n = 1 - \dfrac{1}{n}$ **102.** $a_n = n - \dfrac{1}{n}$

103. $a_n = \dfrac{2^n - 1}{2^n}$ **104.** $a_n = \dfrac{2^n - 1}{3^n}$

105. $a_n = ((-1)^n + 1)\left(\dfrac{n+1}{n}\right)$

106. The first term of a sequence is $x_1 = \cos(1)$. The next terms are $x_2 = x_1$ or $\cos(2)$, whichever is larger; and $x_3 = x_2$ or $\cos(3)$, whichever is larger (farther to the right). In general,

$$x_{n+1} = \max\{x_n, \cos(n+1)\}.$$

107. Nonincreasing sequences A sequence of numbers $\{a_n\}$ in which $a_n \geq a_{n+1}$ for every n is called a **nonincreasing sequence**. A sequence $\{a_n\}$ is **bounded from below** if there is a number M with $M \leq a_n$ for every n. Such a number M is called a **lower bound** for the sequence. Deduce from Theorem 6 that a nonincreasing sequence that is bounded from below converges and that a nonincreasing sequence that is not bounded from below diverges.

(*Continuation of Exercise 107.*) Using the conclusion of Exercise 107, determine which of the sequences in Exercises 108–112 converge and which diverge.

108. $a_n = \dfrac{n+1}{n}$ **109.** $a_n = \dfrac{1 + \sqrt{2n}}{\sqrt{n}}$

110. $a_n = \dfrac{1 - 4^n}{2^n}$ **111.** $a_n = \dfrac{4^{n+1} + 3^n}{4^n}$

112. $a_1 = 1, \quad a_{n+1} = 2a_n - 3$

113. The sequence $\{n/(n+1)\}$ has a least upper bound of 1 Show that if M is a number less than 1, then the terms of $\{n/(n+1)\}$ eventually exceed M. That is, if $M < 1$ there is an integer N such that $n/(n+1) > M$ whenever $n > N$. Since $n/(n+1) < 1$ for every n, this proves that 1 is a least upper bound for $\{n/(n+1)\}$.

114. Uniqueness of least upper bounds Show that if M_1 and M_2 are least upper bounds for the sequence $\{a_n\}$, then $M_1 = M_2$. That is, a sequence cannot have two different least upper bounds.

115. Is it true that a sequence $\{a_n\}$ of positive numbers must converge if it is bounded from above? Give reasons for your answer.

116. Prove that if $\{a_n\}$ is a convergent sequence, then to every positive number ϵ there corresponds an integer N such that for all m and n,

$$m > N \quad \text{and} \quad n > N \quad \Longrightarrow \quad |a_m - a_n| < \epsilon.$$

117. Uniqueness of limits Prove that limits of sequences are unique. That is, show that if L_1 and L_2 are numbers such that $a_n \to L_1$ and $a_n \to L_2$, then $L_1 = L_2$.

118. Limits and subsequences If the terms of one sequence appear in another sequence in their given order, we call the first sequence a **subsequence** of the second. Prove that if two subsequences of a sequence $\{a_n\}$ have different limits $L_1 \neq L_2$, then $\{a_n\}$ diverges.

119. For a sequence $\{a_n\}$ the terms of even index are denoted by a_{2k} and the terms of odd index by a_{2k+1}. Prove that if $a_{2k} \to L$ and $a_{2k+1} \to L$, then $a_n \to L$.

120. Prove that a sequence $\{a_n\}$ converges to 0 if and only if the sequence of absolute values $\{|a_n|\}$ converges to 0.

T Calculator Explorations of Limits

In Exercises 121–124, experiment with a calculator to find a value of N that will make the inequality hold for all $n > N$. Assuming that the inequality is the one from the formal definition of the limit of a sequence, what sequence is being considered in each case and what is its limit?

121. $\left|\sqrt[n]{0.5} - 1\right| < 10^{-3}$ **122.** $\left|\sqrt[n]{n} - 1\right| < 10^{-3}$

123. $(0.9)^n < 10^{-3}$ **124.** $2^n/n! < 10^{-7}$

125. Sequences generated by Newton's method Newton's method, applied to a differentiable function $f(x)$, begins with a starting value x_0 and constructs from it a sequence of numbers $\{x_n\}$ that under favorable circumstances converges to a zero of f. The recursion formula for the sequence is

$$x_{n+1} = x_n - \frac{f(x_n)}{f'(x_n)}.$$

    a. Show that the recursion formula for $f(x) = x^2 - a, a > 0$, can be written as $x_{n+1} = (x_n + a/x_n)/2$.

    b. Starting with $x_0 = 1$ and $a = 3$, calculate successive terms of the sequence until the display begins to repeat. What number is being approximated? Explain.

126. (*Continuation of Exercise 125.*) Repeat part (b) of Exercise 125 with $a = 2$ in place of $a = 3$.

127. A recursive definition of $\pi/2$ If you start with $x_1 = 1$ and define the subsequent terms of $\{x_n\}$ by the rule $x_n = x_{n-1} + \cos x_{n-1}$, you generate a sequence that converges rapidly to $\pi/2$. **a.** Try it. **b.** Use the accompanying figure to explain why the convergence is so rapid.

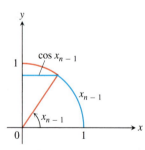

128. According to a front-page article in the December 15, 1992, issue of the *Wall Street Journal*, Ford Motor Company used about $7\frac{1}{4}$ hours of labor to produce stampings for the average vehicle, down from an estimated 15 hours in 1980. The Japanese needed only about $3\frac{1}{2}$ hours.

    Ford's improvement since 1980 represents an average decrease of 6% per year. If that rate continues, then n years from 1992 Ford will use about

$$S_n = 7.25(0.94)^n$$

hours of labor to produce stampings for the average vehicle. Assuming that the Japanese continue to spend $3\frac{1}{2}$ hours per vehicle,

how many more years will it take Ford to catch up? Find out two ways:

    a. Find the first term of the sequence $\{S_n\}$ that is less than or equal to 3.5.

    T b. Graph $f(x) = 7.25(0.94)^x$ and use Trace to find where the graph crosses the line $y = 3.5$.

COMPUTER EXPLORATIONS

Use a CAS to perform the following steps for the sequences in Exercises 129–140.

a. Calculate and then plot the first 25 terms of the sequence. Does the sequence appear to be bounded from above or below? Does it appear to converge or diverge? If it does converge, what is the limit L?

b. If the sequence converges, find an integer N such that $|a_n - L| \leq 0.01$ for $n \geq N$. How far in the sequence do you have to get for the terms to lie within 0.0001 of L?

129. $a_n = \sqrt[n]{n}$ **130.** $a_n = \left(1 + \dfrac{0.5}{n}\right)^n$

131. $a_1 = 1, \quad a_{n+1} = a_n + \dfrac{1}{5^n}$

132. $a_1 = 1, \quad a_{n+1} = a_n + (-2)^n$

133. $a_n = \sin n$ **134.** $a_n = n \sin \dfrac{1}{n}$

135. $a_n = \dfrac{\sin n}{n}$ **136.** $a_n = \dfrac{\ln n}{n}$

137. $a_n = (0.9999)^n$ **138.** $a_n = 123456^{1/n}$

139. $a_n = \dfrac{8^n}{n!}$ **140.** $a_n = \dfrac{n^{41}}{19^n}$

141. Compound interest, deposits, and withdrawals If you invest an amount of money A_0 at a fixed annual interest rate r compounded m times per year, and if the constant amount b is added to the account at the end of each compounding period (or taken from the account if $b < 0$), then the amount you have after $n + 1$ compounding periods is

$$A_{n+1} = \left(1 + \frac{r}{m}\right)A_n + b. \tag{1}$$

    a. If $A_0 = 1000, r = 0.02015, m = 12$, and $b = 50$, calculate and plot the first 100 points (n, A_n). How much money is in your account at the end of 5 years? Does $\{A_n\}$ converge? Is $\{A_n\}$ bounded?

    b. Repeat part (a) with $A_0 = 5000, r = 0.0589, m = 12$, and $b = -50$.

    c. If you invest 5000 dollars in a certificate of deposit (CD) that pays 4.5% annually, compounded quarterly, and you make no further investments in the CD, approximately how many years will it take before you have 20,000 dollars? What if the CD earns 6.25%?

d. It can be shown that for any $k \geq 0$, the sequence defined recursively by Equation (1) satisfies the relation

$$A_k = \left(1 + \frac{r}{m}\right)^k \left(A_0 + \frac{mb}{r}\right) - \frac{mb}{r}. \qquad (2)$$

For the values of the constants A_0, r, m, and b given in part (a), validate this assertion by comparing the values of the first 50 terms of both sequences. Then show by direct substitution that the terms in Equation (2) satisfy the recursion formula in Equation (1).

142. Logistic difference equation The recursive relation

$$a_{n+1} = ra_n(1 - a_n)$$

is called the *logistic difference equation*, and when the initial value a_0 is given the equation defines the *logistic sequence* $\{a_n\}$. Throughout this exercise we choose a_0 in the interval $0 < a_0 < 1$, say $a_0 = 0.3$.

a. Choose $r = 3/4$. Calculate and plot the points (n, a_n) for the first 100 terms in the sequence. Does it appear to converge? What do you guess is the limit? Does the limit seem to depend on your choice of a_0?

b. Choose several values of r in the interval $1 < r < 3$ and repeat the procedures in part (a). Be sure to choose some points near the endpoints of the interval. Describe the behavior of the sequences you observe in your plots.

c. Now examine the behavior of the sequence for values of r near the endpoints of the interval $3 < r < 3.45$. The transition value $r = 3$ is called a **bifurcation value** and the new behavior of the sequence in the interval is called an **attracting 2-cycle**. Explain why this reasonably describes the behavior.

d. Next explore the behavior for r values near the endpoints of each of the intervals $3.45 < r < 3.54$ and $3.54 < r < 3.55$. Plot the first 200 terms of the sequences. Describe in your own words the behavior observed in your plots for each interval. Among how many values does the sequence appear to oscillate for each interval? The values $r = 3.45$ and $r = 3.54$ (rounded to two decimal places) are also called bifurcation values because the behavior of the sequence changes as r crosses over those values.

e. The situation gets even more interesting. There is actually an increasing sequence of bifurcation values $3 < 3.45 < 3.54 < \cdots < c_n < c_{n+1} \cdots$ such that for $c_n < r < c_{n+1}$ the logistic sequence $\{a_n\}$ eventually oscillates steadily among 2^n values, called an **attracting 2^n-cycle**. Moreover, the bifurcation sequence $\{c_n\}$ is bounded above by 3.57 (so it converges). If you choose a value of $r < 3.57$ you will observe a 2^n-cycle of some sort. Choose $r = 3.5695$ and plot 300 points.

f. Let us see what happens when $r > 3.57$. Choose $r = 3.65$ and calculate and plot the first 300 terms of $\{a_n\}$. Observe how the terms wander around in an unpredictable, chaotic fashion. You cannot predict the value of a_{n+1} from previous values of the sequence.

g. For $r = 3.65$ choose two starting values of a_0 that are close together, say, $a_0 = 0.3$ and $a_0 = 0.301$. Calculate and plot the first 300 values of the sequences determined by each starting value. Compare the behaviors observed in your plots. How far out do you go before the corresponding terms of your two sequences appear to depart from each other? Repeat the exploration for $r = 3.75$. Can you see how the plots look different depending on your choice of a_0? We say that the logistic sequence is *sensitive to the initial condition a_0*.

11.2 Infinite Series

An *infinite series* is the sum of an infinite sequence of numbers

$$a_1 + a_2 + a_3 + \cdots + a_n + \cdots$$

The goal of this section is to understand the meaning of such an infinite sum and to develop methods to calculate it. Since there are infinitely many terms to add in an infinite series, we cannot just keep adding to see what comes out. Instead we look at what we get by summing the first n terms of the sequence and stopping. The sum of the first n terms

$$s_n = a_1 + a_2 + a_3 + \cdots + a_n$$

is an ordinary finite sum and can be calculated by normal addition. It is called the *nth partial sum*. As *n* gets larger, we expect the partial sums to get closer and closer to a limiting value in the same sense that the terms of a sequence approach a limit, as discussed in Section 11.1.

For example, to assign meaning to an expression like

$$1 + \frac{1}{2} + \frac{1}{4} + \frac{1}{8} + \frac{1}{16} + \cdots$$

We add the terms one at a time from the beginning and look for a pattern in how these partial sums grow.

Partial sum		Suggestive expression for partial sum	Value
First:	$s_1 = 1$	$2 - 1$	1
Second:	$s_2 = 1 + \frac{1}{2}$	$2 - \frac{1}{2}$	$\frac{3}{2}$
Third:	$s_3 = 1 + \frac{1}{2} + \frac{1}{4}$	$2 - \frac{1}{4}$	$\frac{7}{4}$
$\vdots$	$\vdots$	$\vdots$	$\vdots$
*n*th:	$s_n = 1 + \frac{1}{2} + \frac{1}{4} + \cdots + \frac{1}{2^{n-1}}$	$2 - \frac{1}{2^{n-1}}$	$\frac{2^n - 1}{2^{n-1}}$

Indeed there is a pattern. The partial sums form a sequence whose *n*th term is

$$s_n = 2 - \frac{1}{2^{n-1}}.$$

This sequence of partial sums converges to 2 because $\lim_{n \to \infty}(1/2^n) = 0$. We say

"the sum of the infinite series $1 + \frac{1}{2} + \frac{1}{4} + \cdots + \frac{1}{2^{n-1}} + \cdots$ is 2."

Is the sum of any finite number of terms in this series equal to 2? No. Can we actually add an infinite number of terms one by one? No. But we can still define their sum by defining it to be the limit of the sequence of partial sums as $n \to \infty$, in this case 2 (Figure 11.5). Our knowledge of sequences and limits enables us to break away from the confines of finite sums.

FIGURE 11.5 As the lengths 1, $^1\!/_2$, $^1\!/_4$, $^1\!/_8, \ldots$ are added one by one, the sum approaches 2.

DEFINITIONS Infinite Series, *n*th Term, Partial Sum, Converges, Sum

Given a sequence of numbers $\{a_n\}$, an expression of the form

$$a_1 + a_2 + a_3 + \cdots + a_n + \cdots$$

is an **infinite series**. The number a_n is the ***n*th term** of the series. The sequence $\{s_n\}$ defined by

$$s_1 = a_1$$
$$s_2 = a_1 + a_2$$
$$\vdots$$
$$s_n = a_1 + a_2 + \cdots + a_n = \sum_{k=1}^{n} a_k$$
$$\vdots$$

is the **sequence of partial sums** of the series, the number s_n being the ***n*th partial sum**. If the sequence of partial sums converges to a limit L, we say that the series **converges** and that its **sum** is L. In this case, we also write

$$a_1 + a_2 + \cdots + a_n + \cdots = \sum_{n=1}^{\infty} a_n = L.$$

If the sequence of partial sums of the series does not converge, we say that the series **diverges**.

When we begin to study a given series $a_1 + a_2 + \cdots + a_n + \cdots$, we might not know whether it converges or diverges. In either case, it is convenient to use sigma notation to write the series as

$$\sum_{n=1}^{\infty} a_n, \qquad \sum_{k=1}^{\infty} a_k, \qquad \text{or} \qquad \sum a_n$$

A useful shorthand when summation from 1 to ∞ is understood

Geometric Series

Geometric series are series of the form

$$a + ar + ar^2 + \cdots + ar^{n-1} + \cdots = \sum_{n=1}^{\infty} ar^{n-1}$$

in which a and r are fixed real numbers and $a \neq 0$. The series can also be written as $\sum_{n=0}^{\infty} ar^n$. The **ratio** r can be positive, as in

$$1 + \frac{1}{2} + \frac{1}{4} + \cdots + \left(\frac{1}{2}\right)^{n-1} + \cdots,$$

or negative, as in

$$1 - \frac{1}{3} + \frac{1}{9} - \cdots + \left(-\frac{1}{3}\right)^{n-1} + \cdots.$$

If $r = 1$, the *n*th partial sum of the geometric series is

$$s_n = a + a(1) + a(1)^2 + \cdots + a(1)^{n-1} = na,$$

and the series diverges because $\lim_{n\to\infty} s_n = \pm\infty$, depending on the sign of a. If $r = -1$, the series diverges because the nth partial sums alternate between a and 0. If $|r| \neq 1$, we can determine the convergence or divergence of the series in the following way:

$$s_n = a + ar + ar^2 + \cdots + ar^{n-1}$$

$$rs_n = ar + ar^2 + \cdots + ar^{n-1} + ar^n \qquad \text{Multiply } s_n \text{ by } r.$$

$$s_n - rs_n = a - ar^n \qquad\qquad \text{Subtract } rs_n \text{ from } s_n. \text{ Most of the terms on the right cancel.}$$

$$s_n(1 - r) = a(1 - r^n) \qquad\qquad \text{Factor.}$$

$$s_n = \frac{a(1 - r^n)}{1 - r}, \qquad (r \neq 1). \qquad \text{We can solve for } s_n \text{ if } r \neq 1.$$

If $|r| < 1$, then $r^n \to 0$ as $n \to \infty$ (as in Section 11.1) and $s_n \to a/(1 - r)$. If $|r| > 1$, then $|r^n| \to \infty$ and the series diverges.

If $|r| < 1$, the geometric series $a + ar + ar^2 + \cdots + ar^{n-1} + \cdots$ converges to $a/(1 - r)$:

$$\sum_{n=1}^{\infty} ar^{n-1} = \frac{a}{1 - r}, \qquad |r| < 1.$$

If $|r| \geq 1$, the series diverges.

We have determined when a geometric series converges or diverges, and to what value. Often we can determine that a series converges without knowing the value to which it converges, as we will see in the next several sections. The formula $a/(1 - r)$ for the sum of a geometric series applies *only* when the summation index begins with $n = 1$ in the expression $\sum_{n=1}^{\infty} ar^{n-1}$ (or with the index $n = 0$ if we write the series as $\sum_{n=0}^{\infty} ar^n$).

EXAMPLE 1   Index Starts with $n = 1$

The geometric series with $a = 1/9$ and $r = 1/3$ is

$$\frac{1}{9} + \frac{1}{27} + \frac{1}{81} + \cdots = \sum_{n=1}^{\infty} \frac{1}{9}\left(\frac{1}{3}\right)^{n-1} = \frac{1/9}{1 - (1/3)} = \frac{1}{6}. \qquad \blacksquare$$

EXAMPLE 2   Index Starts with $n = 0$

The series

$$\sum_{n=0}^{\infty} \frac{(-1)^n 5}{4^n} = 5 - \frac{5}{4} + \frac{5}{16} - \frac{5}{64} + \cdots$$

is a geometric series with $a = 5$ and $r = -1/4$. It converges to

$$\frac{a}{1 - r} = \frac{5}{1 + (1/4)} = 4. \qquad \blacksquare$$

EXAMPLE 3   A Bouncing Ball

You drop a ball from a meters above a flat surface. Each time the ball hits the surface after falling a distance h, it rebounds a distance rh, where r is positive but less than 1. Find the total distance the ball travels up and down (Figure 11.6).

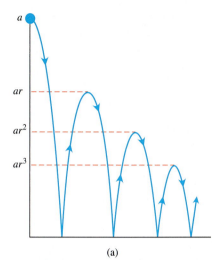

Solution   The total distance is

$$s = a + \underbrace{2ar + 2ar^2 + 2ar^3 + \cdots}_{\text{This sum is } 2ar/(1-r).} = a + \frac{2ar}{1-r} = a\frac{1+r}{1-r}.$$

If $a = 6$ m and $r = 2/3$, for instance, the distance is

$$s = 6\frac{1 + (2/3)}{1 - (2/3)} = 6\left(\frac{5/3}{1/3}\right) = 30 \text{ m.} \quad \blacksquare$$

EXAMPLE 4   Repeating Decimals

Express the repeating decimal $5.232323\ldots$ as the ratio of two integers.

Solution

$$5.232323\ldots = 5 + \frac{23}{100} + \frac{23}{(100)^2} + \frac{23}{(100)^3} + \cdots$$

$$= 5 + \frac{23}{100}\underbrace{\left(1 + \frac{1}{100} + \left(\frac{1}{100}\right)^2 + \cdots\right)}_{1/(1 - 0.01)} \qquad \begin{array}{l} a = 1, \\ r = 1/100 \end{array}$$

$$= 5 + \frac{23}{100}\left(\frac{1}{0.99}\right) = 5 + \frac{23}{99} = \frac{518}{99} \quad \blacksquare$$

Unfortunately, formulas like the one for the sum of a convergent geometric series are rare and we usually have to settle for an estimate of a series' sum (more about this later). The next example, however, is another case in which we can find the sum exactly.

EXAMPLE 5   A Nongeometric but Telescoping Series

Find the sum of the series $\displaystyle\sum_{n=1}^{\infty} \frac{1}{n(n+1)}$.

Solution   We look for a pattern in the sequence of partial sums that might lead to a formula for s_k. The key observation is the partial fraction decomposition

$$\frac{1}{n(n+1)} = \frac{1}{n} - \frac{1}{n+1},$$

so

$$\sum_{n=1}^{k} \frac{1}{n(n+1)} = \sum_{n=1}^{k} \left(\frac{1}{n} - \frac{1}{n+1}\right)$$

and

$$s_k = \left(\frac{1}{1} - \frac{1}{2}\right) + \left(\frac{1}{2} - \frac{1}{3}\right) + \left(\frac{1}{3} - \frac{1}{4}\right) + \cdots + \left(\frac{1}{k} - \frac{1}{k+1}\right).$$

Removing parentheses and canceling adjacent terms of opposite sign collapses the sum to

$$s_k = 1 - \frac{1}{k+1}.$$

FIGURE 11.6   (a) Example 3 shows how to use a geometric series to calculate the total vertical distance traveled by a bouncing ball if the height of each rebound is reduced by the factor r. (b) A stroboscopic photo of a bouncing ball.

We now see that $s_k \rightarrow 1$ as $k \rightarrow \infty$. The series converges, and its sum is 1:

$$\sum_{n=1}^{\infty} \frac{1}{n(n+1)} = 1.$$

Divergent Series

One reason that a series may fail to converge is that its terms don't become small.

EXAMPLE 6 Partial Sums Outgrow Any Number

(a) The series

$$\sum_{n=1}^{\infty} n^2 = 1 + 4 + 9 + \cdots + n^2 + \cdots$$

diverges because the partial sums grow beyond every number L. After $n = 1$, the partial sum $s_n = 1 + 4 + 9 + \cdots + n^2$ is greater than n^2.

(b) The series

$$\sum_{n=1}^{\infty} \frac{n+1}{n} = \frac{2}{1} + \frac{3}{2} + \frac{4}{3} + \cdots + \frac{n+1}{n} + \cdots$$

diverges because the partial sums eventually outgrow every preassigned number. Each term is greater than 1, so the sum of n terms is greater than n.

The nth-Term Test for Divergence

Observe that $\lim_{n \rightarrow \infty} a_n$ must equal zero if the series $\sum_{n=1}^{\infty} a_n$ converges. To see why, let S represent the series' sum and $s_n = a_1 + a_2 + \cdots + a_n$ the nth partial sum. When n is large, both s_n and s_{n-1} are close to S, so their difference, a_n, is close to zero. More formally,

$$a_n = s_n - s_{n-1} \quad \rightarrow \quad S - S = 0. \qquad \text{Difference Rule for sequences}$$

This establishes the following theorem.

Caution
Theorem 7 *does not say* that $\sum_{n=1}^{\infty} a_n$ converges if $a_n \rightarrow 0$. It is possible for a series to diverge when $a_n \rightarrow 0$.

THEOREM 7

If $\sum_{n=1}^{\infty} a_n$ converges, then $a_n \rightarrow 0$.

Theorem 7 leads to a test for detecting the kind of divergence that occurred in Example 6.

The nth-Term Test for Divergence

$\sum_{n=1}^{\infty} a_n$ diverges if $\lim_{n \rightarrow \infty} a_n$ fails to exist or is different from zero.

EXAMPLE 7   Applying the nth-Term Test

(a) $\displaystyle\sum_{n=1}^{\infty} n^2$ diverges because $n^2 \to \infty$

(b) $\displaystyle\sum_{n=1}^{\infty} \frac{n+1}{n}$ diverges because $\dfrac{n+1}{n} \to 1$

(c) $\displaystyle\sum_{n=1}^{\infty} (-1)^{n+1}$ diverges because $\lim_{n\to\infty}(-1)^{n+1}$ does not exist

(d) $\displaystyle\sum_{n=1}^{\infty} \frac{-n}{2n+5}$ diverges because $\lim_{n\to\infty} \dfrac{-n}{2n+5} = -\dfrac{1}{2} \neq 0.$ ∎

EXAMPLE 8 $a_n \to 0$ but the Series Diverges

The series

$$1 + \frac{1}{2} + \frac{1}{2} + \frac{1}{4} + \frac{1}{4} + \frac{1}{4} + \frac{1}{4} + \cdots + \frac{1}{2^n} + \frac{1}{2^n} + \cdots + \frac{1}{2^n} + \cdots$$

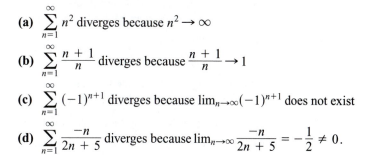

$\underbrace{\qquad}_{\text{2 terms}} \qquad \underbrace{\qquad}_{\text{4 terms}} \qquad \underbrace{\qquad}_{\text{2^n terms}}$

diverges because the terms are grouped into clusters that add to 1, so the partial sums increase without bound. However, the terms of the series form a sequence that converges to 0. Example 1 of Section 11.3 shows that the harmonic series also behaves in this manner.   ∎

Combining Series

Whenever we have two convergent series, we can add them term by term, subtract them term by term, or multiply them by constants to make new convergent series.

THEOREM 8

If $\sum a_n = A$ and $\sum b_n = B$ are convergent series, then

1. *Sum Rule:* $\sum(a_n + b_n) = \sum a_n + \sum b_n = A + B$
2. *Difference Rule:* $\sum(a_n - b_n) = \sum a_n - \sum b_n = A - B$
3. *Constant Multiple Rule:* $\sum ka_n = k\sum a_n = kA$ (Any number k).

Proof   The three rules for series follow from the analogous rules for sequences in Theorem 1, Section 11.1. To prove the Sum Rule for series, let

$$A_n = a_1 + a_2 + \cdots + a_n, \quad B_n = b_1 + b_2 + \cdots + b_n.$$

Then the partial sums of $\sum(a_n + b_n)$ are

$$\begin{aligned}
s_n &= (a_1 + b_1) + (a_2 + b_2) + \cdots + (a_n + b_n)\\
&= (a_1 + \cdots + a_n) + (b_1 + \cdots + b_n)\\
&= A_n + B_n.
\end{aligned}$$

Since $A_n \to A$ and $B_n \to B$, we have $s_n \to A + B$ by the Sum Rule for sequences. The proof of the Difference Rule is similar.

To prove the Constant Multiple Rule for series, observe that the partial sums of $\sum ka_n$ form the sequence

$$s_n = ka_1 + ka_2 + \cdots + ka_n = k(a_1 + a_2 + \cdots + a_n) = kA_n,$$

which converges to kA by the Constant Multiple Rule for sequences. ∎

As corollaries of Theorem 8, we have

1. Every nonzero constant multiple of a divergent series diverges.

2. If $\sum a_n$ converges and $\sum b_n$ diverges, then $\sum(a_n + b_n)$ and $\sum(a_n - b_n)$ both diverge.

We omit the proofs.

CAUTION   Remember that $\sum(a_n + b_n)$ can converge when $\sum a_n$ and $\sum b_n$ both diverge. For example, $\sum a_n = 1 + 1 + 1 + \cdots$ and $\sum b_n = (-1) + (-1) + (-1) + \cdots$ diverge, whereas $\sum(a_n + b_n) = 0 + 0 + 0 + \cdots$ converges to 0.

EXAMPLE 9   Find the sums of the following series.

(a) $\displaystyle\sum_{n=1}^{\infty} \frac{3^{n-1} - 1}{6^{n-1}} = \sum_{n=1}^{\infty} \left(\frac{1}{2^{n-1}} - \frac{1}{6^{n-1}} \right)$

$\qquad\qquad = \displaystyle\sum_{n=1}^{\infty} \frac{1}{2^{n-1}} - \sum_{n=1}^{\infty} \frac{1}{6^{n-1}}$ Difference Rule

$\qquad\qquad = \dfrac{1}{1 - (1/2)} - \dfrac{1}{1 - (1/6)}$ Geometric series with $a = 1$ and $r = 1/2$, $1/6$

$\qquad\qquad = 2 - \dfrac{6}{5}$

$\qquad\qquad = \dfrac{4}{5}$

(b) $\qquad \displaystyle\sum_{n=0}^{\infty} \frac{4}{2^n} = 4\sum_{n=0}^{\infty} \frac{1}{2^n}$ Constant Multiple Rule

$\qquad\qquad\qquad = 4\left(\dfrac{1}{1 - (1/2)} \right)$ Geometric series with $a = 1$, $r = 1/2$

$\qquad\qquad\qquad = 8$ ■

Adding or Deleting Terms

We can add a finite number of terms to a series or delete a finite number of terms without altering the series' convergence or divergence, although in the case of convergence this will usually change the sum. If $\sum_{n=1}^{\infty} a_n$ converges, then $\sum_{n=k}^{\infty} a_n$ converges for any $k > 1$ and

$$\sum_{n=1}^{\infty} a_n = a_1 + a_2 + \cdots + a_{k-1} + \sum_{n=k}^{\infty} a_n.$$

Conversely, if $\sum_{n=k}^{\infty} a_n$ converges for any $k > 1$, then $\sum_{n=1}^{\infty} a_n$ converges. Thus,

$$\sum_{n=1}^{\infty} \frac{1}{5^n} = \frac{1}{5} + \frac{1}{25} + \frac{1}{125} + \sum_{n=4}^{\infty} \frac{1}{5^n}$$

and

$$\sum_{n=4}^{\infty} \frac{1}{5^n} = \left(\sum_{n=1}^{\infty} \frac{1}{5^n} \right) - \frac{1}{5} - \frac{1}{25} - \frac{1}{125}.$$

Reindexing

As long as we preserve the order of its terms, we can reindex any series without altering its convergence. To raise the starting value of the index h units, replace the n in the formula for a_n by $n - h$:

$$\sum_{n=1}^{\infty} a_n = \sum_{n=1+h}^{\infty} a_{n-h} = a_1 + a_2 + a_3 + \cdots.$$

To lower the starting value of the index h units, replace the n in the formula for a_n by $n + h$:

$$\sum_{n=1}^{\infty} a_n = \sum_{n=1-h}^{\infty} a_{n+h} = a_1 + a_2 + a_3 + \cdots.$$

It works like a horizontal shift. We saw this in starting a geometric series with the index $n = 0$ instead of the index $n = 1$, but we can use any other starting index value as well. We usually give preference to indexings that lead to simple expressions.

EXAMPLE 10   Reindexing a Geometric Series

We can write the geometric series

$$\sum_{n=1}^{\infty} \frac{1}{2^{n-1}} = 1 + \frac{1}{2} + \frac{1}{4} + \cdots$$

as

$$\sum_{n=0}^{\infty} \frac{1}{2^n}, \qquad \sum_{n=5}^{\infty} \frac{1}{2^{n-5}}, \qquad \text{or even} \qquad \sum_{n=-4}^{\infty} \frac{1}{2^{n+4}}.$$

The partial sums remain the same no matter what indexing we choose. ∎

EXERCISES 11.2

Finding nth Partial Sums

In Exercises 1–6, find a formula for the nth partial sum of each series and use it to find the series' sum if the series converges.

1. $2 + \frac{2}{3} + \frac{2}{9} + \frac{2}{27} + \cdots + \frac{2}{3^{n-1}} + \cdots$

2. $\frac{9}{100} + \frac{9}{100^2} + \frac{9}{100^3} + \cdots + \frac{9}{100^n} + \cdots$

3. $1 - \frac{1}{2} + \frac{1}{4} - \frac{1}{8} + \cdots + (-1)^{n-1}\frac{1}{2^{n-1}} + \cdots$

4. $1 - 2 + 4 - 8 + \cdots + (-1)^{n-1} 2^{n-1} + \cdots$

5. $\frac{1}{2 \cdot 3} + \frac{1}{3 \cdot 4} + \frac{1}{4 \cdot 5} + \cdots + \frac{1}{(n+1)(n+2)} + \cdots$

6. $\frac{5}{1 \cdot 2} + \frac{5}{2 \cdot 3} + \frac{5}{3 \cdot 4} + \cdots + \frac{5}{n(n+1)} + \cdots$

Series with Geometric Terms

In Exercises 7–14, write out the first few terms of each series to show how the series starts. Then find the sum of the series.

7. $\sum_{n=0}^{\infty} \frac{(-1)^n}{4^n}$

8. $\sum_{n=2}^{\infty} \frac{1}{4^n}$

9. $\displaystyle\sum_{n=1}^{\infty} \frac{7}{4^n}$

10. $\displaystyle\sum_{n=0}^{\infty} (-1)^n \frac{5}{4^n}$

11. $\displaystyle\sum_{n=0}^{\infty} \left(\frac{5}{2^n} + \frac{1}{3^n}\right)$

12. $\displaystyle\sum_{n=0}^{\infty} \left(\frac{5}{2^n} - \frac{1}{3^n}\right)$

13. $\displaystyle\sum_{n=0}^{\infty} \left(\frac{1}{2^n} + \frac{(-1)^n}{5^n}\right)$

14. $\displaystyle\sum_{n=0}^{\infty} \left(\frac{2^{n+1}}{5^n}\right)$

Telescoping Series

Use partial fractions to find the sum of each series in Exercises 15–22.

15. $\displaystyle\sum_{n=1}^{\infty} \frac{4}{(4n-3)(4n+1)}$

16. $\displaystyle\sum_{n=1}^{\infty} \frac{6}{(2n-1)(2n+1)}$

17. $\displaystyle\sum_{n=1}^{\infty} \frac{40n}{(2n-1)^2(2n+1)^2}$

18. $\displaystyle\sum_{n=1}^{\infty} \frac{2n+1}{n^2(n+1)^2}$

19. $\displaystyle\sum_{n=1}^{\infty} \left(\frac{1}{\sqrt{n}} - \frac{1}{\sqrt{n+1}}\right)$

20. $\displaystyle\sum_{n=1}^{\infty} \left(\frac{1}{2^{1/n}} - \frac{1}{2^{1/(n+1)}}\right)$

21. $\displaystyle\sum_{n=1}^{\infty} \left(\frac{1}{\ln(n+2)} - \frac{1}{\ln(n+1)}\right)$

22. $\displaystyle\sum_{n=1}^{\infty} (\tan^{-1}(n) - \tan^{-1}(n+1))$

Convergence or Divergence

Which series in Exercises 23–40 converge, and which diverge? Give reasons for your answers. If a series converges, find its sum.

23. $\displaystyle\sum_{n=0}^{\infty} \left(\frac{1}{\sqrt{2}}\right)^n$

24. $\displaystyle\sum_{n=0}^{\infty} (\sqrt{2})^n$

25. $\displaystyle\sum_{n=1}^{\infty} (-1)^{n+1} \frac{3}{2^n}$

26. $\displaystyle\sum_{n=1}^{\infty} (-1)^{n+1} n$

27. $\displaystyle\sum_{n=0}^{\infty} \cos n\pi$

28. $\displaystyle\sum_{n=0}^{\infty} \frac{\cos n\pi}{5^n}$

29. $\displaystyle\sum_{n=0}^{\infty} e^{-2n}$

30. $\displaystyle\sum_{n=1}^{\infty} \ln \frac{1}{n}$

31. $\displaystyle\sum_{n=1}^{\infty} \frac{2}{10^n}$

32. $\displaystyle\sum_{n=0}^{\infty} \frac{1}{x^n}, \quad |x| > 1$

33. $\displaystyle\sum_{n=0}^{\infty} \frac{2^n - 1}{3^n}$

34. $\displaystyle\sum_{n=1}^{\infty} \left(1 - \frac{1}{n}\right)^n$

35. $\displaystyle\sum_{n=0}^{\infty} \frac{n!}{1000^n}$

36. $\displaystyle\sum_{n=1}^{\infty} \frac{n^n}{n!}$

37. $\displaystyle\sum_{n=1}^{\infty} \ln \left(\frac{n}{n+1}\right)$

38. $\displaystyle\sum_{n=1}^{\infty} \ln \left(\frac{n}{2n+1}\right)$

39. $\displaystyle\sum_{n=0}^{\infty} \left(\frac{e}{\pi}\right)^n$

40. $\displaystyle\sum_{n=0}^{\infty} \frac{e^{n\pi}}{\pi^{ne}}$

Geometric Series

In each of the geometric series in Exercises 41–44, write out the first few terms of the series to find a and r, and find the sum of the series.

Then express the inequality $|r| < 1$ in terms of x and find the values of x for which the inequality holds and the series converges.

41. $\displaystyle\sum_{n=0}^{\infty} (-1)^n x^n$

42. $\displaystyle\sum_{n=0}^{\infty} (-1)^n x^{2n}$

43. $\displaystyle\sum_{n=0}^{\infty} 3\left(\frac{x-1}{2}\right)^n$

44. $\displaystyle\sum_{n=0}^{\infty} \frac{(-1)^n}{2} \left(\frac{1}{3 + \sin x}\right)^n$

In Exercises 45–50, find the values of x for which the given geometric series converges. Also, find the sum of the series (as a function of x) for those values of x.

45. $\displaystyle\sum_{n=0}^{\infty} 2^n x^n$

46. $\displaystyle\sum_{n=0}^{\infty} (-1)^n x^{-2n}$

47. $\displaystyle\sum_{n=0}^{\infty} (-1)^n (x+1)^n$

48. $\displaystyle\sum_{n=0}^{\infty} \left(-\frac{1}{2}\right)^n (x-3)^n$

49. $\displaystyle\sum_{n=0}^{\infty} \sin^n x$

50. $\displaystyle\sum_{n=0}^{\infty} (\ln x)^n$

Repeating Decimals

Express each of the numbers in Exercises 51–58 as the ratio of two integers.

51. $0.\overline{23} = 0.23\ 23\ 23\ldots$

52. $0.\overline{234} = 0.234\ 234\ 234\ldots$

53. $0.\overline{7} = 0.7777\ldots$

54. $0.\overline{d} = 0.dddd\ldots,$ where d is a digit

55. $0.\overline{06} = 0.06666\ldots$

56. $1.\overline{414} = 1.414\ 414\ 414\ldots$

57. $1.24\overline{123} = 1.24\ 123\ 123\ 123\ldots$

58. $3.\overline{142857} = 3.142857\ 142857\ldots$

Theory and Examples

59. The series in Exercise 5 can also be written as

$$\sum_{n=1}^{\infty} \frac{1}{(n+1)(n+2)} \quad \text{and} \quad \sum_{n=-1}^{\infty} \frac{1}{(n+3)(n+4)}.$$

Write it as a sum beginning with **(a)** $n = -2$, **(b)** $n = 0$, **(c)** $n = 5$.

60. The series in Exercise 6 can also be written as

$$\sum_{n=1}^{\infty} \frac{5}{n(n+1)} \quad \text{and} \quad \sum_{n=0}^{\infty} \frac{5}{(n+1)(n+2)}.$$

Write it as a sum beginning with **(a)** $n = -1$, **(b)** $n = 3$, **(c)** $n = 20$.

61. Make up an infinite series of nonzero terms whose sum is

  a. 1   **b.** -3 **c.** 0.

62. (*Continuation of Exercise 61.*) Can you make an infinite series of nonzero terms that converges to any number you want? Explain.

63. Show by example that $\sum (a_n/b_n)$ may diverge even though $\sum a_n$ and $\sum b_n$ converge and no b_n equals 0.

64. Find convergent geometric series $A = \Sigma a_n$ and $B = \Sigma b_n$ that illustrate the fact that $\Sigma a_n b_n$ may converge without being equal to AB.

65. Show by example that $\Sigma (a_n / b_n)$ may converge to something other than A/B even when $A = \Sigma a_n$, $B = \Sigma b_n \neq 0$, and no b_n equals 0.

66. If Σa_n converges and $a_n > 0$ for all n, can anything be said about $\Sigma (1/a_n)$? Give reasons for your answer.

67. What happens if you add a finite number of terms to a divergent series or delete a finite number of terms from a divergent series? Give reasons for your answer.

68. If Σa_n converges and Σb_n diverges, can anything be said about their term-by-term sum $\Sigma (a_n + b_n)$? Give reasons for your answer.

69. Make up a geometric series Σar^{n-1} that converges to the number 5 if

    a. $a = 2$ **b.** $a = 13/2$.

70. Find the value of b for which

$$1 + e^b + e^{2b} + e^{3b} + \cdots = 9.$$

71. For what values of r does the infinite series

$$1 + 2r + r^2 + 2r^3 + r^4 + 2r^5 + r^6 + \cdots$$

converge? Find the sum of the series when it converges.

72. Show that the error $(L - s_n)$ obtained by replacing a convergent geometric series with one of its partial sums s_n is $ar^n/(1 - r)$.

73. A ball is dropped from a height of 4 m. Each time it strikes the pavement after falling from a height of h meters it rebounds to a height of $0.75h$ meters. Find the total distance the ball travels up and down.

74. (*Continuation of Exercise 73.*) Find the total number of seconds the ball in Exercise 73 is traveling. (*Hint:* The formula $s = 4.9t^2$ gives $t = \sqrt{s/4.9}$.)

75. The accompanying figure shows the first five of a sequence of squares. The outermost square has an area of 4 m². Each of the other squares is obtained by joining the midpoints of the sides of the squares before it. Find the sum of the areas of all the squares.

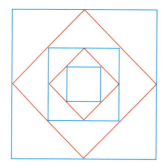

76. The accompanying figure shows the first three rows and part of the fourth row of a sequence of rows of semicircles. There are 2^n semicircles in the nth row, each of radius $1/2^n$. Find the sum of the areas of all the semicircles.

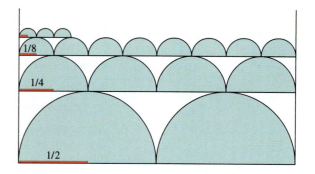

77. Helga von Koch's snowflake curve Helga von Koch's snowflake is a curve of infinite length that encloses a region of finite area. To see why this is so, suppose the curve is generated by starting with an equilateral triangle whose sides have length 1.

    a. Find the length L_n of the nth curve C_n and show that $\lim_{n \to \infty} L_n = \infty$.

    b. Find the area A_n of the region enclosed by C_n and calculate $\lim_{n \to \infty} A_n$.

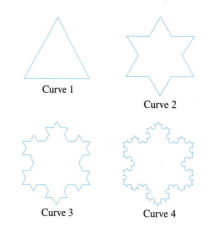

78. The accompanying figure provides an informal proof that $\sum_{n=1}^{\infty} (1/n^2)$ is less than 2. Explain what is going on. (*Source:* "Convergence with Pictures" by P. J. Rippon, *American Mathematical Monthly*, Vol. 93, No. 6, 1986, pp. 476–478.)

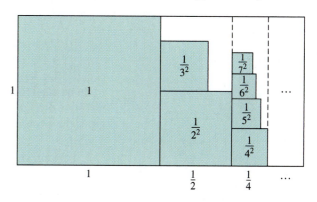

The Integral Test

Given a series $\sum a_n$, we have two questions:

1. Does the series converge?
2. If it converges, what is its sum?

Much of the rest of this chapter is devoted to the first question, and in this section we answer that question by making a connection to the convergence of the improper integral $\int_1^{\infty} f(x)\, dx$. However, as a practical matter the second question is also important, and we will return to it later.

In this section and the next two, we study series that do not have negative terms. The reason for this restriction is that the partial sums of these series form nondecreasing sequences, and nondecreasing sequences that are bounded from above always converge (Theorem 6, Section 11.1). To show that a series of nonnegative terms converges, we need only show that its partial sums are bounded from above.

It may at first seem to be a drawback that this approach establishes the fact of convergence without producing the sum of the series in question. Surely it would be better to compute sums of series directly from formulas for their partial sums. But in most cases such formulas are not available, and in their absence we have to turn instead to the two-step procedure of first establishing convergence and then approximating the sum.

Nondecreasing Partial Sums

Suppose that $\sum_{n=1}^{\infty} a_n$ is an infinite series with $a_n \geq 0$ for all n. Then each partial sum is greater than or equal to its predecessor because $s_{n+1} = s_n + a_n$:

$$s_1 \leq s_2 \leq s_3 \leq \cdots \leq s_n \leq s_{n+1} \leq \cdots.$$

Since the partial sums form a nondecreasing sequence, the Nondecreasing Sequence Theorem (Theorem 6, Section 11.1) tells us that the series will converge if and only if the partial sums are bounded from above.

Corollary of Theorem 6

A series $\sum_{n=1}^{\infty} a_n$ of nonnegative terms converges if and only if its partial sums are bounded from above.

EXAMPLE 1   The Harmonic Series

HISTORICAL BIOGRAPHY

Nicole Oresme
(1320–1382)

The series

$$\sum_{n=1}^{\infty} \frac{1}{n} = 1 + \frac{1}{2} + \frac{1}{3} + \cdots + \frac{1}{n} + \cdots$$

is called the **harmonic series**. The harmonic series is divergent, but this doesn't follow from the nth-Term Test. The nth term $1/n$ does go to zero, but the series still diverges. The reason it diverges is because there is no upper bound for its partial sums. To see why, group the terms of the series in the following way:

$$1 + \frac{1}{2} + \underbrace{\left(\frac{1}{3} + \frac{1}{4}\right)}_{>\frac{2}{4} = \frac{1}{2}} + \underbrace{\left(\frac{1}{5} + \frac{1}{6} + \frac{1}{7} + \frac{1}{8}\right)}_{>\frac{4}{8} = \frac{1}{2}} + \underbrace{\left(\frac{1}{9} + \frac{1}{10} + \cdots + \frac{1}{16}\right)}_{>\frac{8}{16} = \frac{1}{2}} + \cdots.$$

The sum of the first two terms is 1.5. The sum of the next two terms is $1/3 + 1/4$, which is greater than $1/4 + 1/4 = 1/2$. The sum of the next four terms is $1/5 + 1/6 + 1/7 + 1/8$, which is greater than $1/8 + 1/8 + 1/8 + 1/8 = 1/2$. The sum of the next eight terms is $1/9 + 1/10 + 1/11 + 1/12 + 1/13 + 1/14 + 1/15 + 1/16$, which is greater than $8/16 = 1/2$. The sum of the next 16 terms is greater than $16/32 = 1/2$, and so on. In general, the sum of 2^n terms ending with $1/2^{n+1}$ is greater than $2^n/2^{n+1} = 1/2$. The sequence of partial sums is not bounded from above: If $n = 2^k$, the partial sum s_n is greater than $k/2$. The harmonic series diverges. ∎

The Integral Test

We introduce the Integral Test with a series that is related to the harmonic series, but whose nth term is $1/n^2$ instead of $1/n$.

EXAMPLE 2   Does the following series converge?

$$\sum_{n=1}^{\infty} \frac{1}{n^2} = 1 + \frac{1}{4} + \frac{1}{9} + \frac{1}{16} + \cdots + \frac{1}{n^2} + \cdots$$

Solution   We determine the convergence of $\sum_{n=1}^{\infty}(1/n^2)$ by comparing it with $\int_{1}^{\infty}(1/x^2)\,dx$. To carry out the comparison, we think of the terms of the series as values of the function $f(x) = 1/x^2$ and interpret these values as the areas of rectangles under the curve $y = 1/x^2$.

As Figure 11.7 shows,

$$
\begin{aligned}
s_n &= \frac{1}{1^2} + \frac{1}{2^2} + \frac{1}{3^2} + \cdots + \frac{1}{n^2} \\
&= f(1) + f(2) + f(3) + \cdots + f(n) \\
&< f(1) + \int_{1}^{n} \frac{1}{x^2}\,dx \\
&< 1 + \int_{1}^{\infty} \frac{1}{x^2}\,dx \\
&< 1 + 1 = 2.
\end{aligned}
$$

As in Section 8.8, Example 3, $\int_{1}^{\infty}(1/x^2)\,dx = 1$.

Thus the partial sums of $\sum_{n=1}^{\infty} 1/n^2$ are bounded from above (by 2) and the series converges. The sum of the series is known to be $\pi^2/6 \approx 1.64493$. (See Exercise 16 in Section 11.11.) ∎

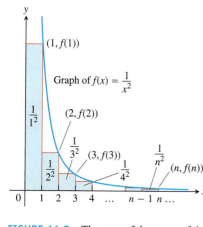

FIGURE 11.7   The sum of the areas of the rectangles under the graph of $f(x) = 1/x^2$ is less than the area under the graph (Example 2).

Caution

The series and integral need not have the same value in the convergent case. As we noted in Example 2, $\sum_{n=1}^{\infty}(1/n^2) = \pi^2/6$ while $\int_{1}^{\infty}(1/x^2)\,dx = 1$.

THEOREM 9   The Integral Test

Let $\{a_n\}$ be a sequence of positive terms. Suppose that $a_n = f(n)$, where f is a continuous, positive, decreasing function of x for all $x \geq N$ (N a positive integer). Then the series $\sum_{n=N}^{\infty} a_n$ and the integral $\int_{N}^{\infty} f(x)\,dx$ both converge or both diverge.

Proof   We establish the test for the case $N = 1$. The proof for general N is similar.

We start with the assumption that f is a decreasing function with $f(n) = a_n$ for every n. This leads us to observe that the rectangles in Figure 11.8a, which have areas

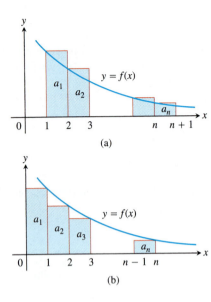

$a_1, a_2, \ldots, a_n$, collectively enclose more area than that under the curve $y = f(x)$ from $x = 1$ to $x = n + 1$. That is,

$$\int_1^{n+1} f(x)\, dx \le a_1 + a_2 + \cdots + a_n.$$

In Figure 11.8b the rectangles have been faced to the left instead of to the right. If we momentarily disregard the first rectangle, of area a_1, we see that

$$a_2 + a_3 + \cdots + a_n \le \int_1^n f(x)\, dx.$$

If we include a_1, we have

$$a_1 + a_2 + \cdots + a_n \le a_1 + \int_1^n f(x)\, dx.$$

Combining these results gives

$$\int_1^{n+1} f(x)\, dx \le a_1 + a_2 + \cdots + a_n \le a_1 + \int_1^n f(x)\, dx.$$

These inequalities hold for each n, and continue to hold as $n \to \infty$.

If $\int_1^\infty f(x)\, dx$ is finite, the right-hand inequality shows that Σa_n is finite. If $\int_1^\infty f(x)\, dx$ is infinite, the left-hand inequality shows that Σa_n is infinite. Hence the series and the integral are both finite or both infinite. ∎

FIGURE 11.8 Subject to the conditions of the Integral Test, the series $\Sigma_{n=1}^\infty a_n$ and the integral $\int_1^\infty f(x)\, dx$ both converge or both diverge.

EXAMPLE 3   The p-Series

Show that the **p-series**

$$\sum_{n=1}^\infty \frac{1}{n^p} = \frac{1}{1^p} + \frac{1}{2^p} + \frac{1}{3^p} + \cdots + \frac{1}{n^p} + \cdots$$

(p a real constant) converges if $p > 1$, and diverges if $p \le 1$.

Solution   If $p > 1$, then $f(x) = 1/x^p$ is a positive decreasing function of x. Since

$$\int_1^\infty \frac{1}{x^p}\, dx = \int_1^\infty x^{-p}\, dx = \lim_{b \to \infty} \left[\frac{x^{-p+1}}{-p + 1} \right]_1^b$$

$$= \frac{1}{1 - p} \lim_{b \to \infty} \left(\frac{1}{b^{p-1}} - 1 \right)$$

$$= \frac{1}{1 - p}(0 - 1) = \frac{1}{p - 1}, \qquad \begin{array}{l} b^{p-1} \to \infty \text{ as } b \to \infty \\ \text{because } p - 1 > 0. \end{array}$$

the series converges by the Integral Test. We emphasize that the sum of the p-series is *not* $1/(p - 1)$. The series converges, but we don't know the value it converges to.

If $p < 1$, then $1 - p > 0$ and

$$\int_1^\infty \frac{1}{x^p}\, dx = \frac{1}{1 - p} \lim_{b \to \infty} (b^{1-p} - 1) = \infty.$$

The series diverges by the Integral Test.

If $p = 1$, we have the (divergent) harmonic series

$$1 + \frac{1}{2} + \frac{1}{3} + \cdots + \frac{1}{n} + \cdots.$$

We have convergence for $p > 1$ but divergence for every other value of p. ∎

The p-series with $p = 1$ is the **harmonic series** (Example 1). The p-Series Test shows that the harmonic series is just *barely* divergent; if we increase p to 1.000000001, for instance, the series converges!

The slowness with which the partial sums of the harmonic series approaches infinity is impressive. For instance, it takes about 178,482,301 terms of the harmonic series to move the partial sums beyond 20. It would take your calculator several weeks to compute a sum with this many terms. (See also Exercise 33b.)

EXAMPLE 4   A Convergent Series

The series

$$\sum_{n=1}^{\infty} \frac{1}{n^2 + 1}$$

converges by the Integral Test. The function $f(x) = 1/(x^2 + 1)$ is positive, continuous, and decreasing for $x \geq 1$, and

$$\int_1^{\infty} \frac{1}{x^2 + 1}\, dx = \lim_{b \to \infty} \left[\arctan x\right]_1^b$$

$$= \lim_{b \to \infty} [\arctan b - \arctan 1]$$

$$= \frac{\pi}{2} - \frac{\pi}{4} = \frac{\pi}{4}.$$

Again we emphasize that $\pi/4$ is *not* the sum of the series. The series converges, but we do not know the value of its sum. ∎

Convergence of the series in Example 4 can also be verified by comparison with the series $\sum 1/n^2$. Comparison tests are studied in the next section.

EXERCISES 11.3

Determining Convergence or Divergence

Which of the series in Exercises 1–30 converge, and which diverge? Give reasons for your answers. (When you check an answer, remember that there may be more than one way to determine the series' convergence or divergence.)

1. $\displaystyle\sum_{n=1}^{\infty} \frac{1}{10^n}$ **2.** $\displaystyle\sum_{n=1}^{\infty} e^{-n}$ **3.** $\displaystyle\sum_{n=1}^{\infty} \frac{n}{n+1}$

4. $\displaystyle\sum_{n=1}^{\infty} \frac{5}{n+1}$ **5.** $\displaystyle\sum_{n=1}^{\infty} \frac{3}{\sqrt{n}}$ **6.** $\displaystyle\sum_{n=1}^{\infty} \frac{-2}{n\sqrt{n}}$

7. $\displaystyle\sum_{n=1}^{\infty} -\frac{1}{8^n}$ **8.** $\displaystyle\sum_{n=1}^{\infty} \frac{-8}{n}$ **9.** $\displaystyle\sum_{n=2}^{\infty} \frac{\ln n}{n}$

10. $\displaystyle\sum_{n=2}^{\infty} \frac{\ln n}{\sqrt{n}}$ **11.** $\displaystyle\sum_{n=1}^{\infty} \frac{2^n}{3^n}$ **12.** $\displaystyle\sum_{n=1}^{\infty} \frac{5^n}{4^n + 3}$

13. $\displaystyle\sum_{n=0}^{\infty} \frac{-2}{n+1}$ **14.** $\displaystyle\sum_{n=1}^{\infty} \frac{1}{2n-1}$ **15.** $\displaystyle\sum_{n=1}^{\infty} \frac{2^n}{n+1}$

16. $\displaystyle\sum_{n=1}^{\infty} \frac{1}{\sqrt{n}(\sqrt{n}+1)}$ **17.** $\displaystyle\sum_{n=2}^{\infty} \frac{\sqrt{n}}{\ln n}$ **18.** $\displaystyle\sum_{n=1}^{\infty} \left(1 + \frac{1}{n}\right)^n$

19. $\displaystyle\sum_{n=1}^{\infty} \frac{1}{(\ln 2)^n}$ **20.** $\displaystyle\sum_{n=1}^{\infty} \frac{1}{(\ln 3)^n}$

21. $\displaystyle\sum_{n=3}^{\infty} \frac{(1/n)}{(\ln n)\sqrt{\ln^2 n - 1}}$ **22.** $\displaystyle\sum_{n=1}^{\infty} \frac{1}{n(1 + \ln^2 n)}$

23. $\sum_{n=1}^{\infty} n \sin \frac{1}{n}$

24. $\sum_{n=1}^{\infty} n \tan \frac{1}{n}$

25. $\sum_{n=1}^{\infty} \frac{e^n}{1 + e^{2n}}$

26. $\sum_{n=1}^{\infty} \frac{2}{1 + e^n}$

27. $\sum_{n=1}^{\infty} \frac{8 \tan^{-1} n}{1 + n^2}$

28. $\sum_{n=1}^{\infty} \frac{n}{n^2 + 1}$

29. $\sum_{n=1}^{\infty} \operatorname{sech} n$

30. $\sum_{n=1}^{\infty} \operatorname{sech}^2 n$

Theory and Examples

For what values of a, if any, do the series in Exercises 31 and 32 converge?

31. $\sum_{n=1}^{\infty} \left(\frac{a}{n+2} - \frac{1}{n+4} \right)$

32. $\sum_{n=3}^{\infty} \left(\frac{1}{n-1} - \frac{2a}{n+1} \right)$

33. a. Draw illustrations like those in Figures 11.7 and 11.8 to show that the partial sums of the harmonic series satisfy the inequalities

$$\ln(n+1) = \int_1^{n+1} \frac{1}{x} \, dx \le 1 + \frac{1}{2} + \cdots + \frac{1}{n}$$

$$\le 1 + \int_1^n \frac{1}{x} \, dx = 1 + \ln n.$$

T **b.** There is absolutely no empirical evidence for the divergence of the harmonic series even though we know it diverges. The partial sums just grow too slowly. To see what we mean, suppose you had started with $s_1 = 1$ the day the universe was formed, 13 billion years ago, and added a new term every *second*. About how large would the partial sum s_n be today, assuming a 365-day year?

34. Are there any values of x for which $\sum_{n=1}^{\infty} (1/(nx))$ converges? Give reasons for your answer.

35. Is it true that if $\sum_{n=1}^{\infty} a_n$ is a divergent series of positive numbers then there is also a divergent series $\sum_{n=1}^{\infty} b_n$ of positive numbers with $b_n < a_n$ for every n? Is there a "smallest" divergent series of positive numbers? Give reasons for your answers.

36. (*Continuation of Exercise 35.*) Is there a "largest" convergent series of positive numbers? Explain.

37. The Cauchy condensation test The Cauchy condensation test says: Let $\{a_n\}$ be a nonincreasing sequence ($a_n \ge a_{n+1}$ for all n) of positive terms that converges to 0. Then $\sum a_n$ converges if and only if $\sum 2^n a_{2^n}$ converges. For example, $\sum (1/n)$ diverges because $\sum 2^n \cdot (1/2^n) = \sum 1$ diverges. Show why the test works.

38. Use the Cauchy condensation test from Exercise 37 to show that

a. $\sum_{n=2}^{\infty} \frac{1}{n \ln n}$ diverges;

b. $\sum_{n=1}^{\infty} \frac{1}{n^p}$ converges if $p > 1$ and diverges if $p \le 1$.

39. Logarithmic *p*-series

a. Show that

$$\int_2^{\infty} \frac{dx}{x(\ln x)^p} \quad (p \text{ a positive constant})$$

converges if and only if $p > 1$.

b. What implications does the fact in part (a) have for the convergence of the series

$$\sum_{n=2}^{\infty} \frac{1}{n(\ln n)^p} ?$$

Give reasons for your answer.

40. (*Continuation of Exercise 39.*) Use the result in Exercise 39 to determine which of the following series converge and which diverge. Support your answer in each case.

a. $\sum_{n=2}^{\infty} \frac{1}{n(\ln n)}$

b. $\sum_{n=2}^{\infty} \frac{1}{n(\ln n)^{1.01}}$

c. $\sum_{n=2}^{\infty} \frac{1}{n \ln (n^3)}$

d. $\sum_{n=2}^{\infty} \frac{1}{n(\ln n)^3}$

41. Euler's constant Graphs like those in Figure 11.8 suggest that as n increases there is little change in the difference between the sum

$$1 + \frac{1}{2} + \cdots + \frac{1}{n}$$

and the integral

$$\ln n = \int_1^n \frac{1}{x} \, dx.$$

To explore this idea, carry out the following steps.

a. By taking $f(x) = 1/x$ in the proof of Theorem 9, show that

$$\ln(n+1) \le 1 + \frac{1}{2} + \cdots + \frac{1}{n} \le 1 + \ln n$$

or

$$0 < \ln(n+1) - \ln n \le 1 + \frac{1}{2} + \cdots + \frac{1}{n} - \ln n \le 1.$$

Thus, the sequence

$$a_n = 1 + \frac{1}{2} + \cdots + \frac{1}{n} - \ln n$$

is bounded from below and from above.

b. Show that

$$\frac{1}{n+1} < \int_n^{n+1} \frac{1}{x} \, dx = \ln(n+1) - \ln n,$$

and use this result to show that the sequence $\{a_n\}$ in part (a) is decreasing.

Since a decreasing sequence that is bounded from below converges (Exercise 107 in Section 11.1), the numbers a_n defined in part (a) converge:

$$1 + \frac{1}{2} + \cdots + \frac{1}{n} - \ln n \to \gamma.$$

The number γ, whose value is $0.5772\ldots$, is called *Euler's constant*. In contrast to other special numbers like π and e, no other expression with a simple law of formulation has ever been found for γ.

42. Use the integral test to show that

$$\sum_{n=0}^{\infty} e^{-n^2}$$

converges.

11.4 Comparison Tests

We have seen how to determine the convergence of geometric series, *p*-series, and a few others. We can test the convergence of many more series by comparing their terms to those of a series whose convergence is known.

THEOREM 10   The Comparison Test

Let $\sum a_n$ be a series with no negative terms.

(a) $\sum a_n$ converges if there is a convergent series $\sum c_n$ with $a_n \leq c_n$ for all $n > N$, for some integer N.

(b) $\sum a_n$ diverges if there is a divergent series of nonnegative terms $\sum d_n$ with $a_n \geq d_n$ for all $n > N$, for some integer N.

Proof   In Part (a), the partial sums of $\sum a_n$ are bounded above by

$$M = a_1 + a_2 + \cdots + a_N + \sum_{n=N+1}^{\infty} c_n.$$

They therefore form a nondecreasing sequence with a limit $L \leq M$.

In Part (b), the partial sums of $\sum a_n$ are not bounded from above. If they were, the partial sums for $\sum d_n$ would be bounded by

$$M^* = d_1 + d_2 + \cdots + d_N + \sum_{n=N+1}^{\infty} a_n$$

and $\sum d_n$ would have to converge instead of diverge.   ∎

EXAMPLE 1   Applying the Comparison Test

(a) The series

$$\sum_{n=1}^{\infty} \frac{5}{5n - 1}$$

diverges because its *n*th term

$$\frac{5}{5n - 1} = \frac{1}{n - \frac{1}{5}} > \frac{1}{n}$$

is greater than the *n*th term of the divergent harmonic series.

(b) The series

$$\sum_{n=0}^{\infty} \frac{1}{n!} = 1 + \frac{1}{1!} + \frac{1}{2!} + \frac{1}{3!} + \cdots$$

converges because its terms are all positive and less than or equal to the corresponding terms of

$$1 + \sum_{n=0}^{\infty} \frac{1}{2^n} = 1 + 1 + \frac{1}{2} + \frac{1}{2^2} + \cdots.$$

The geometric series on the left converges and we have

$$1 + \sum_{n=0}^{\infty} \frac{1}{2^n} = 1 + \frac{1}{1 - (1/2)} = 3.$$

The fact that 3 is an upper bound for the partial sums of $\sum_{n=0}^{\infty} (1/n!)$ does not mean that the series converges to 3. As we will see in Section 11.9, the series converges to e.

(c) The series

$$5 + \frac{2}{3} + \frac{1}{7} + 1 + \frac{1}{2 + \sqrt{1}} + \frac{1}{4 + \sqrt{2}} + \frac{1}{8 + \sqrt{3}} + \cdots + \frac{1}{2^n + \sqrt{n}} + \cdots$$

converges. To see this, we ignore the first three terms and compare the remaining terms with those of the convergent geometric series $\sum_{n=0}^{\infty} (1/2^n)$. The term $1/(2^n + \sqrt{n})$ of the truncated sequence is less than the corresponding term $1/2^n$ of the geometric series. We see that term by term we have the comparison,

$$1 + \frac{1}{2 + \sqrt{1}} + \frac{1}{4 + \sqrt{2}} + \frac{1}{8 + \sqrt{3}} + \cdots \leq 1 + \frac{1}{2} + \frac{1}{4} + \frac{1}{8} + \cdots$$

So the truncated series and the original series converge by an application of the Comparison Test. ∎

The Limit Comparison Test

We now introduce a comparison test that is particularly useful for series in which a_n is a rational function of n.

THEOREM 11   Limit Comparison Test

Suppose that $a_n > 0$ and $b_n > 0$ for all $n \geq N$ (N an integer).

1. If $\lim\limits_{n \to \infty} \dfrac{a_n}{b_n} = c > 0$, then $\sum a_n$ and $\sum b_n$ both converge or both diverge.

2. If $\lim\limits_{n \to \infty} \dfrac{a_n}{b_n} = 0$ and $\sum b_n$ converges, then $\sum a_n$ converges.

3. If $\lim\limits_{n \to \infty} \dfrac{a_n}{b_n} = \infty$ and $\sum b_n$ diverges, then $\sum a_n$ diverges.

Proof   We will prove Part 1. Parts 2 and 3 are left as Exercises 37(a) and (b). Since $c/2 > 0$, there exists an integer N such that for all n

$$n > N \implies \left| \frac{a_n}{b_n} - c \right| < \frac{c}{2}.$$

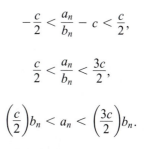

Limit definition with $\epsilon = c/2$, $L = c$, and a_n replaced by a_n/b_n

Thus, for $n > N$,

$$-\frac{c}{2} < \frac{a_n}{b_n} - c < \frac{c}{2},$$

$$\frac{c}{2} < \frac{a_n}{b_n} < \frac{3c}{2},$$

$$\left(\frac{c}{2} \right) b_n < a_n < \left(\frac{3c}{2} \right) b_n.$$

If $\sum b_n$ converges, then $\sum (3c/2) b_n$ converges and $\sum a_n$ converges by the Direct Comparison Test. If $\sum b_n$ diverges, then $\sum (c/2) b_n$ diverges and $\sum a_n$ diverges by the Direct Comparison Test.   ∎

EXAMPLE 2   Using the Limit Comparison Test

Which of the following series converge, and which diverge?

(a) $\dfrac{3}{4} + \dfrac{5}{9} + \dfrac{7}{16} + \dfrac{9}{25} + \cdots = \displaystyle\sum_{n=1}^{\infty} \dfrac{2n + 1}{(n + 1)^2} = \sum_{n=1}^{\infty} \dfrac{2n + 1}{n^2 + 2n + 1}$

(b) $\dfrac{1}{1} + \dfrac{1}{3} + \dfrac{1}{7} + \dfrac{1}{15} + \cdots = \displaystyle\sum_{n=1}^{\infty} \dfrac{1}{2^n - 1}$

(c) $\dfrac{1 + 2 \ln 2}{9} + \dfrac{1 + 3 \ln 3}{14} + \dfrac{1 + 4 \ln 4}{21} + \cdots = \displaystyle\sum_{n=2}^{\infty} \dfrac{1 + n \ln n}{n^2 + 5}$

Solution

(a)   Let $a_n = (2n + 1)/(n^2 + 2n + 1)$. For large n, we expect a_n to behave like $2n/n^2 = 2/n$ since the leading terms dominate for large n, so we let $b_n = 1/n$. Since

$$\sum_{n=1}^{\infty} b_n = \sum_{n=1}^{\infty} \frac{1}{n} \text{ diverges}$$

and

$$\lim_{n \to \infty} \frac{a_n}{b_n} = \lim_{n \to \infty} \frac{2n^2 + n}{n^2 + 2n + 1} = 2,$$

$\sum a_n$ diverges by Part 1 of the Limit Comparison Test. We could just as well have taken $b_n = 2/n$, but $1/n$ is simpler.

(b) Let $a_n = 1/(2^n - 1)$. For large n, we expect a_n to behave like $1/2^n$, so we let $b_n = 1/2^n$. Since

$$\sum_{n=1}^{\infty} b_n = \sum_{n=1}^{\infty} \frac{1}{2^n} \quad \text{converges}$$

and

$$\lim_{n \to \infty} \frac{a_n}{b_n} = \lim_{n \to \infty} \frac{2^n}{2^n - 1}$$

$$= \lim_{n \to \infty} \frac{1}{1 - (1/2^n)}$$

$$= 1,$$

$\sum a_n$ converges by Part 1 of the Limit Comparison Test.

(c) Let $a_n = (1 + n \ln n)/(n^2 + 5)$. For large n, we expect a_n to behave like $(n \ln n)/n^2 = (\ln n)/n$, which is greater than $1/n$ for $n \geq 3$, so we take $b_n = 1/n$. Since

$$\sum_{n=2}^{\infty} b_n = \sum_{n=2}^{\infty} \frac{1}{n} \quad \text{diverges}$$

and

$$\lim_{n \to \infty} \frac{a_n}{b_n} = \lim_{n \to \infty} \frac{n + n^2 \ln n}{n^2 + 5}$$

$$= \infty,$$

$\sum a_n$ diverges by Part 3 of the Limit Comparison Test. ∎

EXAMPLE 3 Does $\displaystyle\sum_{n=1}^{\infty} \frac{\ln n}{n^{3/2}}$ converge?

Solution Because $\ln n$ grows more slowly than n^c for any positive constant c (Section 11.1, Exercise 91), we would expect to have

$$\frac{\ln n}{n^{3/2}} < \frac{n^{1/4}}{n^{3/2}} = \frac{1}{n^{5/4}}$$

for n sufficiently large. Indeed, taking $a_n = (\ln n)/n^{3/2}$ and $b_n = 1/n^{5/4}$, we have

$$\lim_{n \to \infty} \frac{a_n}{b_n} = \lim_{n \to \infty} \frac{\ln n}{n^{1/4}}$$

$$= \lim_{n \to \infty} \frac{1/n}{(1/4)n^{-3/4}} \qquad \text{l'Hôpital's Rule}$$

$$= \lim_{n \to \infty} \frac{4}{n^{1/4}} = 0.$$

Since $\sum b_n = \sum(1/n^{5/4})$ (a p-series with $p > 1$) converges, $\sum a_n$ converges by Part 2 of the Limit Comparison Test. ∎

EXERCISES 11.4

Determining Convergence or Divergence

Which of the series in Exercises 1–36 converge, and which diverge?
Give reasons for your answers.

1. $\displaystyle\sum_{n=1}^{\infty} \frac{1}{2\sqrt{n} + \sqrt[3]{n}}$

2. $\displaystyle\sum_{n=1}^{\infty} \frac{3}{n + \sqrt{n}}$

3. $\displaystyle\sum_{n=1}^{\infty} \frac{\sin^2 n}{2^n}$

4. $\displaystyle\sum_{n=1}^{\infty} \frac{1 + \cos n}{n^2}$

5. $\displaystyle\sum_{n=1}^{\infty} \frac{2n}{3n - 1}$

6. $\displaystyle\sum_{n=1}^{\infty} \frac{n + 1}{n^2\sqrt{n}}$

7. $\displaystyle\sum_{n=1}^{\infty} \left(\frac{n}{3n + 1}\right)^n$

8. $\displaystyle\sum_{n=1}^{\infty} \frac{1}{\sqrt{n^3 + 2}}$

9. $\displaystyle\sum_{n=3}^{\infty} \frac{1}{\ln(\ln n)}$

10. $\displaystyle\sum_{n=2}^{\infty} \frac{1}{(\ln n)^2}$

11. $\displaystyle\sum_{n=1}^{\infty} \frac{(\ln n)^2}{n^3}$

12. $\displaystyle\sum_{n=1}^{\infty} \frac{(\ln n)^3}{n^3}$

13. $\displaystyle\sum_{n=2}^{\infty} \frac{1}{\sqrt{n}\,\ln n}$

14. $\displaystyle\sum_{n=1}^{\infty} \frac{(\ln n)^2}{n^{3/2}}$

15. $\displaystyle\sum_{n=1}^{\infty} \frac{1}{1 + \ln n}$

16. $\displaystyle\sum_{n=1}^{\infty} \frac{1}{(1 + \ln n)^2}$

17. $\displaystyle\sum_{n=2}^{\infty} \frac{\ln(n + 1)}{n + 1}$

18. $\displaystyle\sum_{n=1}^{\infty} \frac{1}{(1 + \ln^2 n)}$

19. $\displaystyle\sum_{n=2}^{\infty} \frac{1}{n\sqrt{n^2 - 1}}$

20. $\displaystyle\sum_{n=1}^{\infty} \frac{\sqrt{n}}{n^2 + 1}$

21. $\displaystyle\sum_{n=1}^{\infty} \frac{1 - n}{n2^n}$

22. $\displaystyle\sum_{n=1}^{\infty} \frac{n + 2^n}{n^2 2^n}$

23. $\displaystyle\sum_{n=1}^{\infty} \frac{1}{3^{n-1} + 1}$

24. $\displaystyle\sum_{n=1}^{\infty} \frac{3^{n-1} + 1}{3^n}$

25. $\displaystyle\sum_{n=1}^{\infty} \sin\frac{1}{n}$

26. $\displaystyle\sum_{n=1}^{\infty} \tan\frac{1}{n}$

27. $\displaystyle\sum_{n=1}^{\infty} \frac{10n + 1}{n(n + 1)(n + 2)}$

28. $\displaystyle\sum_{n=3}^{\infty} \frac{5n^3 - 3n}{n^2(n - 2)(n^2 + 5)}$

29. $\displaystyle\sum_{n=1}^{\infty} \frac{\tan^{-1} n}{n^{1.1}}$

30. $\displaystyle\sum_{n=1}^{\infty} \frac{\sec^{-1} n}{n^{1.3}}$

31. $\displaystyle\sum_{n=1}^{\infty} \frac{\coth n}{n^2}$

32. $\displaystyle\sum_{n=1}^{\infty} \frac{\tanh n}{n^2}$

33. $\displaystyle\sum_{n=1}^{\infty} \frac{1}{n\sqrt[n]{n}}$

34. $\displaystyle\sum_{n=1}^{\infty} \frac{\sqrt[n]{n}}{n^2}$

35. $\displaystyle\sum_{n=1}^{\infty} \frac{1}{1 + 2 + 3 + \cdots + n}$

36. $\displaystyle\sum_{n=1}^{\infty} \frac{1}{1 + 2^2 + 3^2 + \cdots + n^2}$

Theory and Examples

37. Prove **(a)** Part 2 and **(b)** Part 3 of the Limit Comparison Test.

38. If $\sum_{n=1}^{\infty} a_n$ is a convergent series of nonnegative numbers, can anything be said about $\sum_{n=1}^{\infty}(a_n/n)$? Explain.

39. Suppose that $a_n > 0$ and $b_n > 0$ for $n \geq N$ (N an integer). If $\lim_{n\to\infty}(a_n/b_n) = \infty$ and $\sum a_n$ converges, can anything be said about $\sum b_n$? Give reasons for your answer.

40. Prove that if $\sum a_n$ is a convergent series of nonnegative terms, then $\sum a_n^2$ converges.

COMPUTER EXPLORATION

41. It is not yet known whether the series

$$\sum_{n=1}^{\infty} \frac{1}{n^3 \sin^2 n}$$

converges or diverges. Use a CAS to explore the behavior of the series by performing the following steps.

a. Define the sequence of partial sums

$$s_k = \sum_{n=1}^{k} \frac{1}{n^3 \sin^2 n}.$$

What happens when you try to find the limit of s_k as $k \to \infty$? Does your CAS find a closed form answer for this limit?

b. Plot the first 100 points (k, s_k) for the sequence of partial sums. Do they appear to converge? What would you estimate the limit to be?

c. Next plot the first 200 points (k, s_k). Discuss the behavior in your own words.

d. Plot the first 400 points (k, s_k). What happens when $k = 355$? Calculate the number $355/113$. Explain from your calculation what happened at $k = 355$. For what values of k would you guess this behavior might occur again?

You will find an interesting discussion of this series in Chapter 72 of *Mazes for the Mind* by Clifford A. Pickover, St. Martin's Press, Inc., New York, 1992.

11.5 The Ratio and Root Tests

The Ratio Test measures the rate of growth (or decline) of a series by examining the ratio a_{n+1}/a_n. For a geometric series $\sum ar^n$, this rate is a constant $((ar^{n+1})/(ar^n) = r)$, and the series converges if and only if its ratio is less than 1 in absolute value. The Ratio Test is a powerful rule extending that result. We prove it on the next page using the Comparison Test.

THEOREM 12   The Ratio Test

Let $\sum a_n$ be a series with positive terms and suppose that

$$\lim_{n \to \infty} \frac{a_{n+1}}{a_n} = \rho.$$

Then

(a) the series *converges* if $\rho < 1$,

(b) the series *diverges* if $\rho > 1$ or ρ is infinite,

(c) the test is *inconclusive* if $\rho = 1$.

Proof

(a) $\boldsymbol{\rho < 1.}$ Let r be a number between ρ and 1. Then the number $\epsilon = r - \rho$ is positive. Since

$$\frac{a_{n+1}}{a_n} \to \rho,$$

a_{n+1}/a_n must lie within ϵ of ρ when n is large enough, say for all $n \geq N$. In particular

$$\frac{a_{n+1}}{a_n} < \rho + \epsilon = r, \qquad \text{when } n \geq N.$$

That is,

$$a_{N+1} < ra_N,$$
$$a_{N+2} < ra_{N+1} < r^2 a_N,$$
$$a_{N+3} < ra_{N+2} < r^3 a_N,$$
$$\vdots$$
$$a_{N+m} < ra_{N+m-1} < r^m a_N.$$

These inequalities show that the terms of our series, after the Nth term, approach zero more rapidly than the terms in a geometric series with ratio $r < 1$. More precisely, consider the series $\sum c_n$, where $c_n = a_n$ for $n = 1, 2, \ldots, N$ and $c_{N+1} = ra_N$, $c_{N+2} = r^2 a_N, \ldots, c_{N+m} = r^m a_N, \ldots$. Now $a_n \leq c_n$ for all n, and

$$\sum_{n=1}^{\infty} c_n = a_1 + a_2 + \cdots + a_{N-1} + a_N + ra_N + r^2 a_N + \cdots$$
$$= a_1 + a_2 + \cdots + a_{N-1} + a_N(1 + r + r^2 + \cdots).$$

The geometric series $1 + r + r^2 + \cdots$ converges because $|r| < 1$, so $\sum c_n$ converges. Since $a_n \leq c_n$, $\sum a_n$ also converges.

(b) $\boldsymbol{1 < \rho \leq \infty.}$ From some index M on,

$$\frac{a_{n+1}}{a_n} > 1 \qquad \text{and} \qquad a_M < a_{M+1} < a_{M+2} < \cdots.$$

The terms of the series do not approach zero as n becomes infinite, and the series diverges by the nth-Term Test.

(c) $\rho = 1.$ The two series

$$\sum_{n=1}^{\infty} \frac{1}{n} \quad \text{and} \quad \sum_{n=1}^{\infty} \frac{1}{n^2}$$

show that some other test for convergence must be used when $\rho = 1$.

$$\text{For } \sum_{n=1}^{\infty} \frac{1}{n}: \quad \frac{a_{n+1}}{a_n} = \frac{1/(n+1)}{1/n} = \frac{n}{n+1} \to 1.$$

$$\text{For } \sum_{n=1}^{\infty} \frac{1}{n^2}: \quad \frac{a_{n+1}}{a_n} = \frac{1/(n+1)^2}{1/n^2} = \left(\frac{n}{n+1}\right)^2 \to 1^2 = 1.$$

In both cases, $\rho = 1$, yet the first series diverges, whereas the second converges.   ∎

The Ratio Test is often effective when the terms of a series contain factorials of expressions involving n or expressions raised to a power involving n.

EXAMPLE 1   Applying the Ratio Test

Investigate the convergence of the following series.

(a) $\displaystyle\sum_{n=0}^{\infty} \frac{2^n + 5}{3^n}$ **(b)** $\displaystyle\sum_{n=1}^{\infty} \frac{(2n)!}{n!n!}$ **(c)** $\displaystyle\sum_{n=1}^{\infty} \frac{4^n n!n!}{(2n)!}$

Solution

(a) For the series $\sum_{n=0}^{\infty} (2^n + 5)/3^n$,

$$\frac{a_{n+1}}{a_n} = \frac{(2^{n+1} + 5)/3^{n+1}}{(2^n + 5)/3^n} = \frac{1}{3} \cdot \frac{2^{n+1} + 5}{2^n + 5} = \frac{1}{3} \cdot \left(\frac{2 + 5 \cdot 2^{-n}}{1 + 5 \cdot 2^{-n}}\right) \to \frac{1}{3} \cdot \frac{2}{1} = \frac{2}{3}.$$

The series converges because $\rho = 2/3$ is less than 1. This does *not* mean that 2/3 is the sum of the series. In fact,

$$\sum_{n=0}^{\infty} \frac{2^n + 5}{3^n} = \sum_{n=0}^{\infty} \left(\frac{2}{3}\right)^n + \sum_{n=0}^{\infty} \frac{5}{3^n} = \frac{1}{1 - (2/3)} + \frac{5}{1 - (1/3)} = \frac{21}{2}.$$

(b) If $a_n = \dfrac{(2n)!}{n!n!}$, then $a_{n+1} = \dfrac{(2n+2)!}{(n+1)!(n+1)!}$ and

$$\frac{a_{n+1}}{a_n} = \frac{n!n!(2n+2)(2n+1)(2n)!}{(n+1)!(n+1)!(2n)!}$$

$$= \frac{(2n+2)(2n+1)}{(n+1)(n+1)} = \frac{4n+2}{n+1} \to 4.$$

The series diverges because $\rho = 4$ is greater than 1.

(c) If $a_n = 4^n n!n!/(2n)!$, then

$$\frac{a_{n+1}}{a_n} = \frac{4^{n+1}(n+1)!(n+1)!}{(2n+2)(2n+1)(2n)!} \cdot \frac{(2n)!}{4^n n!n!}$$

$$= \frac{4(n+1)(n+1)}{(2n+2)(2n+1)} = \frac{2(n+1)}{2n+1} \to 1.$$

Because the limit is $\rho = 1$, we cannot decide from the Ratio Test whether the series converges. When we notice that $a_{n+1}/a_n = (2n + 2)/(2n + 1)$, we conclude that a_{n+1} is always greater than a_n because $(2n + 2)/(2n + 1)$ is always greater than 1. Therefore, all terms are greater than or equal to $a_1 = 2$, and the nth term does not approach zero as $n \to \infty$. The series diverges. ∎

The Root Test

The convergence tests we have so far for Σa_n work best when the formula for a_n is relatively simple. But consider the following.

EXAMPLE 2 Let $a_n = \begin{cases} n/2^n, & n \text{ odd} \\ 1/2^n, & n \text{ even.} \end{cases}$ Does Σa_n converge?

Solution We write out several terms of the series:

$$\sum_{n=1}^{\infty} a_n = \frac{1}{2^1} + \frac{1}{2^2} + \frac{3}{2^3} + \frac{1}{2^4} + \frac{5}{2^5} + \frac{1}{2^6} + \frac{7}{2^7} + \cdots$$

$$= \frac{1}{2} + \frac{1}{4} + \frac{3}{8} + \frac{1}{16} + \frac{5}{32} + \frac{1}{64} + \frac{7}{128} + \cdots.$$

Clearly, this is not a geometric series. The nth term approaches zero as $n \to \infty$, so we do not know if the series diverges. The Integral Test does not look promising. The Ratio Test produces

$$\frac{a_{n+1}}{a_n} = \begin{cases} \dfrac{1}{2n}, & n \text{ odd} \\ \dfrac{n + 1}{2}, & n \text{ even.} \end{cases}$$

As $n \to \infty$, the ratio is alternately small and large and has no limit. A test that will answer the question (the series converges) is the Root Test. ∎

THEOREM 13   The Root Test

Let Σa_n be a series with $a_n \geq 0$ for $n \geq N$, and suppose that

$$\lim_{n \to \infty} \sqrt[n]{a_n} = \rho.$$

Then

(a) the series *converges* if $\rho < 1$,

(b) the series *diverges* if $\rho > 1$ or ρ is infinite,

(c) the test is *inconclusive* if $\rho = 1$.

Proof

(a) $\rho < 1$. Choose an $\epsilon > 0$ so small that $\rho + \epsilon < 1$. Since $\sqrt[n]{a_n} \to \rho$, the terms $\sqrt[n]{a_n}$ eventually get closer than ϵ to ρ. In other words, there exists an index $M \geq N$ such that

$$\sqrt[n]{a_n} < \rho + \epsilon \qquad \text{when } n \geq M.$$

Then it is also true that

$$a_n < (\rho + \epsilon)^n \qquad \text{for } n \geq M.$$

Now, $\sum_{n=M}^{\infty} (\rho + \epsilon)^n$, a geometric series with ratio $(\rho + \epsilon) < 1$, converges. By comparison, $\sum_{n=M}^{\infty} a_n$ converges, from which it follows that

$$\sum_{n=1}^{\infty} a_n = a_1 + \cdots + a_{M-1} + \sum_{n=M}^{\infty} a_n$$

converges.

(b) $1 < \rho \leq \infty$. For all indices beyond some integer M, we have $\sqrt[n]{a_n} > 1$, so that $a_n > 1$ for $n > M$. The terms of the series do not converge to zero. The series diverges by the nth-Term Test.

(c) $\rho = 1$. The series $\sum_{n=1}^{\infty} (1/n)$ and $\sum_{n=1}^{\infty} (1/n^2)$ show that the test is not conclusive when $\rho = 1$. The first series diverges and the second converges, but in both cases $\sqrt[n]{a_n} \to 1$. ∎

EXAMPLE 3   Applying the Root Test

Which of the following series converges, and which diverges?

(a) $\sum_{n=1}^{\infty} \dfrac{n^2}{2^n}$ (b) $\sum_{n=1}^{\infty} \dfrac{2^n}{n^2}$ (c) $\sum_{n=1}^{\infty} \left(\dfrac{1}{1+n}\right)^n$

Solution

(a) $\sum_{n=1}^{\infty} \dfrac{n^2}{2^n}$ converges because $\sqrt[n]{\dfrac{n^2}{2^n}} = \dfrac{\sqrt[n]{n^2}}{\sqrt[n]{2^n}} = \dfrac{\left(\sqrt[n]{n}\right)^2}{2} \to \dfrac{1}{2} < 1$.

(b) $\sum_{n=1}^{\infty} \dfrac{2^n}{n^2}$ diverges because $\sqrt[n]{\dfrac{2^n}{n^2}} = \dfrac{2}{\left(\sqrt[n]{n}\right)^2} \to \dfrac{2}{1} > 1$.

(c) $\sum_{n=1}^{\infty} \left(\dfrac{1}{1+n}\right)^n$ converges because $\sqrt[n]{\left(\dfrac{1}{1+n}\right)^n} = \dfrac{1}{1+n} \to 0 < 1$. ∎

EXAMPLE 2   Revisited

Let $a_n = \begin{cases} n/2^n, & n \text{ odd} \\ 1/2^n, & n \text{ even.} \end{cases}$ Does $\sum a_n$ converge?

Solution   We apply the Root Test, finding that

$$\sqrt[n]{a_n} = \begin{cases} \sqrt[n]{n}/2, & n \text{ odd} \\ 1/2, & n \text{ even.} \end{cases}$$

Therefore,

$$\frac{1}{2} \leq \sqrt[n]{a_n} \leq \frac{\sqrt[n]{n}}{2}.$$

Since $\sqrt[n]{n} \to 1$ (Section 11.1, Theorem 5), we have $\lim_{n \to \infty} \sqrt[n]{a_n} = 1/2$ by the Sandwich Theorem. The limit is less than 1, so the series converges by the Root Test. ∎

EXERCISES 11.5

Determining Convergence or Divergence

Which of the series in Exercises 1–26 converge, and which diverge? Give reasons for your answers. (When checking your answers, remember there may be more than one way to determine a series' convergence or divergence.)

1. $\displaystyle\sum_{n=1}^{\infty} \frac{n^{\sqrt{2}}}{2^n}$

2. $\displaystyle\sum_{n=1}^{\infty} n^2 e^{-n}$

3. $\displaystyle\sum_{n=1}^{\infty} n! e^{-n}$

4. $\displaystyle\sum_{n=1}^{\infty} \frac{n!}{10^n}$

5. $\displaystyle\sum_{n=1}^{\infty} \frac{n^{10}}{10^n}$

6. $\displaystyle\sum_{n=1}^{\infty} \left(\frac{n-2}{n}\right)^n$

7. $\displaystyle\sum_{n=1}^{\infty} \frac{2 + (-1)^n}{1.25^n}$

8. $\displaystyle\sum_{n=1}^{\infty} \frac{(-2)^n}{3^n}$

9. $\displaystyle\sum_{n=1}^{\infty} \left(1 - \frac{3}{n}\right)^n$

10. $\displaystyle\sum_{n=1}^{\infty} \left(1 - \frac{1}{3n}\right)^n$

11. $\displaystyle\sum_{n=1}^{\infty} \frac{\ln n}{n^3}$

12. $\displaystyle\sum_{n=1}^{\infty} \frac{(\ln n)^n}{n^n}$

13. $\displaystyle\sum_{n=1}^{\infty} \left(\frac{1}{n} - \frac{1}{n^2}\right)$

14. $\displaystyle\sum_{n=1}^{\infty} \left(\frac{1}{n} - \frac{1}{n^2}\right)^n$

15. $\displaystyle\sum_{n=1}^{\infty} \frac{\ln n}{n}$

16. $\displaystyle\sum_{n=1}^{\infty} \frac{n \ln n}{2^n}$

17. $\displaystyle\sum_{n=1}^{\infty} \frac{(n+1)(n+2)}{n!}$

18. $\displaystyle\sum_{n=1}^{\infty} e^{-n}(n^3)$

19. $\displaystyle\sum_{n=1}^{\infty} \frac{(n+3)!}{3! n! 3^n}$

20. $\displaystyle\sum_{n=1}^{\infty} \frac{n 2^n (n+1)!}{3^n n!}$

21. $\displaystyle\sum_{n=1}^{\infty} \frac{n!}{(2n+1)!}$

22. $\displaystyle\sum_{n=1}^{\infty} \frac{n!}{n^n}$

23. $\displaystyle\sum_{n=2}^{\infty} \frac{n}{(\ln n)^n}$

24. $\displaystyle\sum_{n=2}^{\infty} \frac{n}{(\ln n)^{(n/2)}}$

25. $\displaystyle\sum_{n=1}^{\infty} \frac{n! \ln n}{n(n+2)!}$

26. $\displaystyle\sum_{n=1}^{\infty} \frac{3^n}{n^3 2^n}$

Which of the series $\sum_{n=1}^{\infty} a_n$ defined by the formulas in Exercises 27–38 converge, and which diverge? Give reasons for your answers.

27. $a_1 = 2, \quad a_{n+1} = \dfrac{1 + \sin n}{n} a_n$

28. $a_1 = 1, \quad a_{n+1} = \dfrac{1 + \tan^{-1} n}{n} a_n$

29. $a_1 = \dfrac{1}{3}, \quad a_{n+1} = \dfrac{3n - 1}{2n + 5} a_n$

30. $a_1 = 3, \quad a_{n+1} = \dfrac{n}{n + 1} a_n$

31. $a_1 = 2, \quad a_{n+1} = \dfrac{2}{n} a_n$

32. $a_1 = 5, \quad a_{n+1} = \dfrac{\sqrt[n]{n}}{2} a_n$

33. $a_1 = 1, \quad a_{n+1} = \dfrac{1 + \ln n}{n} a_n$

34. $a_1 = \dfrac{1}{2}, \quad a_{n+1} = \dfrac{n + \ln n}{n + 10} a_n$

35. $a_1 = \dfrac{1}{3}, \quad a_{n+1} = \sqrt[n]{a_n}$

36. $a_1 = \dfrac{1}{2}, \quad a_{n+1} = (a_n)^{n+1}$

37. $a_n = \dfrac{2^n n! n!}{(2n)!}$

38. $a_n = \dfrac{(3n)!}{n!(n + 1)!(n + 2)!}$

Which of the series in Exercises 39–44 converge, and which diverge? Give reasons for your answers.

39. $\displaystyle\sum_{n=1}^{\infty} \frac{(n!)^n}{(n^n)^2}$

40. $\displaystyle\sum_{n=1}^{\infty} \frac{(n!)^n}{n^{(n^2)}}$

41. $\displaystyle\sum_{n=1}^{\infty} \frac{n^n}{2^{(n^2)}}$

42. $\displaystyle\sum_{n=1}^{\infty} \frac{n^n}{(2^n)^2}$

43. $\displaystyle\sum_{n=1}^{\infty} \frac{1 \cdot 3 \cdot \cdots \cdot (2n - 1)}{4^n 2^n n!}$

44. $\displaystyle\sum_{n=1}^{\infty} \frac{1 \cdot 3 \cdot \cdots \cdot (2n - 1)}{[2 \cdot 4 \cdot \cdots \cdot (2n)](3^n + 1)}$

Theory and Examples

45. Neither the Ratio nor the Root Test helps with p-series. Try them on

$$\sum_{n=1}^{\infty} \frac{1}{n^p}$$

and show that both tests fail to provide information about convergence.

46. Show that neither the Ratio Test nor the Root Test provides information about the convergence of

$$\sum_{n=2}^{\infty} \frac{1}{(\ln n)^p} \qquad (p \text{ constant}).$$

47. Let $a_n = \begin{cases} n/2^n, & \text{if } n \text{ is a prime number} \\ 1/2^n, & \text{otherwise.} \end{cases}$
Does $\sum a_n$ converge? Give reasons for your answer.

11.6 Alternating Series, Absolute and Conditional Convergence

A series in which the terms are alternately positive and negative is an **alternating series**. Here are three examples:

$$1 - \frac{1}{2} + \frac{1}{3} - \frac{1}{4} + \frac{1}{5} - \cdots + \frac{(-1)^{n+1}}{n} + \cdots \tag{1}$$

$$-2 + 1 - \frac{1}{2} + \frac{1}{4} - \frac{1}{8} + \cdots + \frac{(-1)^{n}4}{2^{n}} + \cdots \tag{2}$$

$$1 - 2 + 3 - 4 + 5 - 6 + \cdots + (-1)^{n+1}n + \cdots \tag{3}$$

Series (1), called the **alternating harmonic series**, converges, as we will see in a moment. Series (2) a geometric series with ratio $r = -1/2$, converges to $-2/[1 + (1/2)] = -4/3$. Series (3) diverges because the nth term does not approach zero.

We prove the convergence of the alternating harmonic series by applying the Alternating Series Test.

THEOREM 14   **The Alternating Series Test (Leibniz's Theorem)**

The series

$$\sum_{n=1}^{\infty}(-1)^{n+1}u_n = u_1 - u_2 + u_3 - u_4 + \cdots$$

converges if all three of the following conditions are satisfied:

1. The u_n's are all positive.
2. $u_n \geq u_{n+1}$ for all $n \geq N$, for some integer N.
3. $u_n \to 0$.

Proof   If n is an even integer, say $n = 2m$, then the sum of the first n terms is

$$s_{2m} = (u_1 - u_2) + (u_3 - u_4) + \cdots + (u_{2m-1} - u_{2m})$$
$$= u_1 - (u_2 - u_3) - (u_4 - u_5) - \cdots - (u_{2m-2} - u_{2m-1}) - u_{2m}.$$

The first equality shows that s_{2m} is the sum of m nonnegative terms, since each term in parentheses is positive or zero. Hence $s_{2m+2} \geq s_{2m}$, and the sequence $\{s_{2m}\}$ is nondecreasing. The second equality shows that $s_{2m} \leq u_1$. Since $\{s_{2m}\}$ is nondecreasing and bounded from above, it has a limit, say

$$\lim_{m \to \infty} s_{2m} = L. \tag{4}$$

If n is an odd integer, say $n = 2m + 1$, then the sum of the first n terms is $s_{2m+1} = s_{2m} + u_{2m+1}$. Since $u_n \to 0$,

$$\lim_{m \to \infty} u_{2m+1} = 0$$

and, as $m \to \infty$,

$$s_{2m+1} = s_{2m} + u_{2m+1} \to L + 0 = L. \tag{5}$$

Combining the results of Equations (4) and (5) gives $\lim_{n \to \infty} s_n = L$ (Section 11.1, Exercise 119). ∎

EXAMPLE 1   The alternating harmonic series

$$\sum_{n=1}^{\infty}(-1)^{n+1}\frac{1}{n} = 1 - \frac{1}{2} + \frac{1}{3} - \frac{1}{4} + \cdots$$

satisfies the three requirements of Theorem 14 with $N = 1$; it therefore converges.   ∎

FIGURE 11.9   The partial sums of an alternating series that satisfies the hypotheses of Theorem 14 for $N = 1$ straddle the limit from the beginning.

A graphical interpretation of the partial sums (Figure 11.9) shows how an alternating series converges to its limit L when the three conditions of Theorem 14 are satisfied with $N = 1$. (Exercise 63 asks you to picture the case $N > 1$.) Starting from the origin of the x-axis, we lay off the positive distance $s_1 = u_1$. To find the point corresponding to $s_2 = u_1 - u_2$, we back up a distance equal to u_2. Since $u_2 \leq u_1$, we do not back up any farther than the origin. We continue in this seesaw fashion, backing up or going forward as the signs in the series demand. But for $n \geq N$, each forward or backward step is shorter than (or at most the same size as) the preceding step, because $u_{n+1} \leq u_n$. And since the nth term approaches zero as n increases, the size of step we take forward or backward gets smaller and smaller. We oscillate across the limit L, and the amplitude of oscillation approaches zero. The limit L lies between any two successive sums s_n and s_{n+1} and hence differs from s_n by an amount less than u_{n+1}.

Because

$$|L - s_n| < u_{n+1} \qquad \text{for } n \geq N,$$

we can make useful estimates of the sums of convergent alternating series.

> **THEOREM 15   The Alternating Series Estimation Theorem**
> If the alternating series $\sum_{n=1}^{\infty}(-1)^{n+1}u_n$ satisfies the three conditions of Theorem 14, then for $n \geq N$,
>
> $$s_n = u_1 - u_2 + \cdots + (-1)^{n+1}u_n$$
>
> approximates the sum L of the series with an error whose absolute value is less than u_{n+1}, the numerical value of the first unused term. Furthermore, the remainder, $L - s_n$, has the same sign as the first unused term.

We leave the verification of the sign of the remainder for Exercise 53.

EXAMPLE 2   We try Theorem 15 on a series whose sum we know:

$$\sum_{n=0}^{\infty}(-1)^n\frac{1}{2^n} = 1 - \frac{1}{2} + \frac{1}{4} - \frac{1}{8} + \frac{1}{16} - \frac{1}{32} + \frac{1}{64} - \frac{1}{128} \;\vdots\; + \frac{1}{256} - \cdots.$$

The theorem says that if we truncate the series after the eighth term, we throw away a total that is positive and less than $1/256$. The sum of the first eight terms is 0.6640625. The sum of the series is

$$\frac{1}{1 - (-1/2)} = \frac{1}{3/2} = \frac{2}{3}.$$

The difference, $(2/3) - 0.6640625 = 0.0026041666\ldots$, is positive and less than $(1/256) = 0.00390625$.   ∎

Absolute and Conditional Convergence

> **DEFINITION**   **Absolutely Convergent**
>
> A series $\sum a_n$ **converges absolutely** (is **absolutely convergent**) if the corresponding series of absolute values, $\sum |a_n|$, converges.

The geometric series

$$1 - \frac{1}{2} + \frac{1}{4} - \frac{1}{8} + \cdots$$

converges absolutely because the corresponding series of absolute values

$$1 + \frac{1}{2} + \frac{1}{4} + \frac{1}{8} + \cdots$$

converges. The alternating harmonic series does not converge absolutely. The corresponding series of absolute values is the (divergent) harmonic series.

> **DEFINITION**   **Conditionally Convergent**
>
> A series that converges but does not converge absolutely **converges conditionally**.

The alternating harmonic series converges conditionally.

Absolute convergence is important for two reasons. First, we have good tests for convergence of series of positive terms. Second, if a series converges absolutely, then it converges. That is the thrust of the next theorem.

> **THEOREM 16**   **The Absolute Convergence Test**
>
> If $\displaystyle\sum_{n=1}^{\infty} |a_n|$ converges, then $\displaystyle\sum_{n=1}^{\infty} a_n$ converges.

Proof   For each n,

$$-|a_n| \le a_n \le |a_n|, \qquad \text{so} \qquad 0 \le a_n + |a_n| \le 2|a_n|.$$

If $\sum_{n=1}^{\infty} |a_n|$ converges, then $\sum_{n=1}^{\infty} 2|a_n|$ converges and, by the Direct Comparison Test, the nonnegative series $\sum_{n=1}^{\infty} (a_n + |a_n|)$ converges. The equality $a_n = (a_n + |a_n|) - |a_n|$ now lets us express $\sum_{n=1}^{\infty} a_n$ as the difference of two convergent series:

$$\sum_{n=1}^{\infty} a_n = \sum_{n=1}^{\infty} (a_n + |a_n| - |a_n|) = \sum_{n=1}^{\infty} (a_n + |a_n|) - \sum_{n=1}^{\infty} |a_n|.$$

Therefore, $\sum_{n=1}^{\infty} a_n$ converges. ∎

CAUTION   We can rephrase Theorem 16 to say that every absolutely convergent series converges. However, the converse statement is false: Many convergent series do not converge absolutely (such as the alternating harmonic series in Example 1).

EXAMPLE 3   Applying the Absolute Convergence Test

(a) For $\sum_{n=1}^{\infty}(-1)^{n+1}\dfrac{1}{n^2} = 1 - \dfrac{1}{4} + \dfrac{1}{9} - \dfrac{1}{16} + \cdots$, the corresponding series of absolute values is the convergent series

$$\sum_{n=1}^{\infty}\frac{1}{n^2} = 1 + \frac{1}{4} + \frac{1}{9} + \frac{1}{16} + \cdots.$$

The original series converges because it converges absolutely.

(b) For $\sum_{n=1}^{\infty}\dfrac{\sin n}{n^2} = \dfrac{\sin 1}{1} + \dfrac{\sin 2}{4} + \dfrac{\sin 3}{9} + \cdots$, the corresponding series of absolute values is

$$\sum_{n=1}^{\infty}\left|\frac{\sin n}{n^2}\right| = \frac{|\sin 1|}{1} + \frac{|\sin 2|}{4} + \cdots,$$

which converges by comparison with $\sum_{n=1}^{\infty}(1/n^2)$ because $|\sin n| \le 1$ for every n. The original series converges absolutely; therefore it converges. ∎

EXAMPLE 4   Alternating p-Series

If p is a positive constant, the sequence $\{1/n^p\}$ is a decreasing sequence with limit zero. Therefore the alternating p-series

$$\sum_{n=1}^{\infty}\frac{(-1)^{n-1}}{n^p} = 1 - \frac{1}{2^p} + \frac{1}{3^p} - \frac{1}{4^p} + \cdots, \qquad p > 0$$

converges.

If $p > 1$, the series converges absolutely. If $0 < p \le 1$, the series converges conditionally.

Conditional convergence: $\quad 1 - \dfrac{1}{\sqrt{2}} + \dfrac{1}{\sqrt{3}} - \dfrac{1}{\sqrt{4}} + \cdots$

Absolute convergence: $\quad 1 - \dfrac{1}{2^{3/2}} + \dfrac{1}{3^{3/2}} - \dfrac{1}{4^{3/2}} + \cdots$ ∎

Rearranging Series

> **THEOREM 17**   **The Rearrangement Theorem for Absolutely Convergent Series**
>
> If $\sum_{n=1}^{\infty} a_n$ converges absolutely, and $b_1, b_2, \ldots, b_n, \ldots$ is any arrangement of the sequence $\{a_n\}$, then $\sum b_n$ converges absolutely and
>
> $$\sum_{n=1}^{\infty} b_n = \sum_{n=1}^{\infty} a_n.$$

(For an outline of the proof, see Exercise 60.)

EXAMPLE 5 Applying the Rearrangement Theorem

As we saw in Example 3, the series

$$1 - \frac{1}{4} + \frac{1}{9} - \frac{1}{16} + \cdots + (-1)^{n-1}\frac{1}{n^2} + \cdots$$

converges absolutely. A possible rearrangement of the terms of the series might start with a positive term, then two negative terms, then three positive terms, then four negative terms, and so on: After k terms of one sign, take $k + 1$ terms of the opposite sign. The first ten terms of such a series look like this:

$$1 - \frac{1}{4} - \frac{1}{16} + \frac{1}{9} + \frac{1}{25} + \frac{1}{49} - \frac{1}{36} - \frac{1}{64} - \frac{1}{100} - \frac{1}{144} + \cdots.$$

The Rearrangement Theorem says that both series converge to the same value. In this example, if we had the second series to begin with, we would probably be glad to exchange it for the first, if we knew that we could. We can do even better: The sum of either series is also equal to

$$\sum_{n=1}^{\infty}\frac{1}{(2n-1)^2} - \sum_{n=1}^{\infty}\frac{1}{(2n)^2}.$$

(See Exercise 61.) ∎

If we rearrange infinitely many terms of a conditionally convergent series, we can get results that are far different from the sum of the original series. Here is an example.

EXAMPLE 6 Rearranging the Alternating Harmonic Series

The alternating harmonic series

$$\frac{1}{1} - \frac{1}{2} + \frac{1}{3} - \frac{1}{4} + \frac{1}{5} - \frac{1}{6} + \frac{1}{7} - \frac{1}{8} + \frac{1}{9} - \frac{1}{10} + \frac{1}{11} - \cdots$$

can be rearranged to diverge or to reach any preassigned sum.

(a) *Rearranging* $\sum_{n=1}^{\infty}(-1)^{n+1}/n$ *to diverge.* The series of terms $\sum[1/(2n-1)]$ diverges to $+\infty$ and the series of terms $\sum(-1/2n)$ diverges to $-\infty$. No matter how far out in the sequence of odd-numbered terms we begin, we can always add enough positive terms to get an arbitrarily large sum. Similarly, with the negative terms, no matter how far out we start, we can add enough consecutive even-numbered terms to get a negative sum of arbitrarily large absolute value. If we wished to do so, we could start adding odd-numbered terms until we had a sum greater than $+3$, say, and then follow that with enough consecutive negative terms to make the new total less than -4. We could then add enough positive terms to make the total greater than $+5$ and follow with consecutive unused negative terms to make a new total less than -6, and so on. In this way, we could make the swings arbitrarily large in either direction.

(b) *Rearranging* $\sum_{n=1}^{\infty}(-1)^{n+1}/n$ *to converge to 1.* Another possibility is to focus on a particular limit. Suppose we try to get sums that converge to 1. We start with the first term, $1/1$, and then subtract $1/2$. Next we add $1/3$ and $1/5$, which brings the total back to 1 or above. Then we add consecutive negative terms until the total is less than 1. We continue in this manner: When the sum is less than 1, add positive terms until the total is 1 or more; then subtract (add negative) terms until the total is again less than 1. This process can be continued indefinitely. Because both the odd-numbered terms and the even-numbered terms of the original series approach zero as $n \to \infty$, the amount by which our partial sums exceed 1 or fall below it approaches zero. So the new series converges to 1. The rearranged series starts like this:

$$\frac{1}{1} - \frac{1}{2} + \frac{1}{3} + \frac{1}{5} - \frac{1}{4} + \frac{1}{7} + \frac{1}{9} - \frac{1}{6} + \frac{1}{11} + \frac{1}{13} - \frac{1}{8} + \frac{1}{15} + \frac{1}{17} - \frac{1}{10}$$

$$+ \frac{1}{19} + \frac{1}{21} - \frac{1}{12} + \frac{1}{23} + \frac{1}{25} - \frac{1}{14} + \frac{1}{27} - \frac{1}{16} + \cdots$$ ∎

The kind of behavior illustrated by the series in Example 6 is typical of what can happen with any conditionally convergent series. Therefore we must always add the terms of a conditionally convergent series in the order given.

We have now developed several tests for convergence and divergence of series. In summary:

---

1. **The *n*th-Term Test:** Unless $a_n \to 0$, the series diverges.
2. **Geometric series:** $\sum ar^n$ converges if $|r| < 1$; otherwise it diverges.
3. **p-series:** $\sum 1/n^p$ converges if $p > 1$; otherwise it diverges.
4. **Series with nonnegative terms:** Try the Integral Test, Ratio Test, or Root Test. Try comparing to a known series with the Comparison Test.
5. **Series with some negative terms:** Does $\sum |a_n|$ converge? If yes, so does $\sum a_n$, since absolute convergence implies convergence.
6. **Alternating series:** $\sum a_n$ converges if the series satisfies the conditions of the Alternating Series Test.

EXERCISES 11.6

Determining Convergence or Divergence

Which of the alternating series in Exercises 1–10 converge, and which diverge? Give reasons for your answers.

1. $\sum_{n=1}^{\infty} (-1)^{n+1} \frac{1}{n^2}$

2. $\sum_{n=1}^{\infty} (-1)^{n+1} \frac{1}{n^{3/2}}$

3. $\sum_{n=1}^{\infty} (-1)^{n+1} \left(\frac{n}{10}\right)^n$

4. $\sum_{n=1}^{\infty} (-1)^{n+1} \frac{10^n}{n^{10}}$

5. $\sum_{n=2}^{\infty} (-1)^{n+1} \frac{1}{\ln n}$

6. $\sum_{n=1}^{\infty} (-1)^{n+1} \frac{\ln n}{n}$

7. $\sum_{n=2}^{\infty} (-1)^{n+1} \frac{\ln n}{\ln n^2}$

8. $\sum_{n=1}^{\infty} (-1)^n \ln\left(1 + \frac{1}{n}\right)$

9. $\sum_{n=1}^{\infty} (-1)^{n+1} \frac{\sqrt{n}+1}{n+1}$

10. $\sum_{n=1}^{\infty} (-1)^{n+1} \frac{3\sqrt{n}+1}{\sqrt{n}+1}$

Absolute Convergence

Which of the series in Exercises 11–44 converge absolutely, which converge, and which diverge? Give reasons for your answers.

11. $\sum_{n=1}^{\infty} (-1)^{n+1} (0.1)^n$

12. $\sum_{n=1}^{\infty} (-1)^{n+1} \frac{(0.1)^n}{n}$

13. $\sum_{n=1}^{\infty} (-1)^n \frac{1}{\sqrt{n}}$

14. $\sum_{n=1}^{\infty} \frac{(-1)^n}{1 + \sqrt{n}}$

15. $\sum_{n=1}^{\infty} (-1)^{n+1} \frac{n}{n^3 + 1}$

16. $\sum_{n=1}^{\infty} (-1)^{n+1} \frac{n!}{2^n}$

17. $\sum_{n=1}^{\infty} (-1)^n \frac{1}{n + 3}$

18. $\sum_{n=1}^{\infty} (-1)^n \frac{\sin n}{n^2}$

19. $\sum_{n=1}^{\infty} (-1)^{n+1} \frac{3 + n}{5 + n}$

20. $\sum_{n=2}^{\infty} (-1)^n \frac{1}{\ln (n^3)}$

21. $\sum_{n=1}^{\infty} (-1)^{n+1} \frac{1 + n}{n^2}$

22. $\sum_{n=1}^{\infty} \frac{(-2)^{n+1}}{n + 5^n}$

23. $\sum_{n=1}^{\infty} (-1)^n n^2 (2/3)^n$

24. $\sum_{n=1}^{\infty} (-1)^{n+1} \left(\sqrt[n]{10}\right)$

25. $\sum_{n=1}^{\infty} (-1)^n \frac{\tan^{-1} n}{n^2 + 1}$

26. $\sum_{n=2}^{\infty} (-1)^{n+1} \frac{1}{n \ln n}$

27. $\displaystyle\sum_{n=1}^{\infty}(-1)^{n}\frac{n}{n+1}$

28. $\displaystyle\sum_{n=1}^{\infty}(-1)^{n}\frac{\ln n}{n-\ln n}$

29. $\displaystyle\sum_{n=1}^{\infty}\frac{(-100)^{n}}{n!}$

30. $\displaystyle\sum_{n=1}^{\infty}(-5)^{-n}$

31. $\displaystyle\sum_{n=1}^{\infty}\frac{(-1)^{n-1}}{n^{2}+2n+1}$

32. $\displaystyle\sum_{n=2}^{\infty}(-1)^{n}\left(\frac{\ln n}{\ln n^{2}}\right)^{n}$

33. $\displaystyle\sum_{n=1}^{\infty}\frac{\cos n\pi}{n\sqrt{n}}$

34. $\displaystyle\sum_{n=1}^{\infty}\frac{\cos n\pi}{n}$

35. $\displaystyle\sum_{n=1}^{\infty}\frac{(-1)^{n}(n+1)^{n}}{(2n)^{n}}$

36. $\displaystyle\sum_{n=1}^{\infty}\frac{(-1)^{n+1}(n!)^{2}}{(2n)!}$

37. $\displaystyle\sum_{n=1}^{\infty}(-1)^{n}\frac{(2n)!}{2^{n}n!n}$

38. $\displaystyle\sum_{n=1}^{\infty}(-1)^{n}\frac{(n!)^{2}3^{n}}{(2n+1)!}$

39. $\displaystyle\sum_{n=1}^{\infty}(-1)^{n}\left(\sqrt{n+1}-\sqrt{n}\right)$ **40.** $\displaystyle\sum_{n=1}^{\infty}(-1)^{n}\left(\sqrt{n^{2}+n}-n\right)$

41. $\displaystyle\sum_{n=1}^{\infty}(-1)^{n}\left(\sqrt{n+\sqrt{n}}-\sqrt{n}\right)$

42. $\displaystyle\sum_{n=1}^{\infty}\frac{(-1)^{n}}{\sqrt{n}+\sqrt{n+1}}$

43. $\displaystyle\sum_{n=1}^{\infty}(-1)^{n}\operatorname{sech}n$

44. $\displaystyle\sum_{n=1}^{\infty}(-1)^{n}\operatorname{csch}n$

Error Estimation

In Exercises 45–48, estimate the magnitude of the error involved in using the sum of the first four terms to approximate the sum of the entire series.

45. $\displaystyle\sum_{n=1}^{\infty}(-1)^{n+1}\frac{1}{n}$ It can be shown that the sum is $\ln 2$.

46. $\displaystyle\sum_{n=1}^{\infty}(-1)^{n+1}\frac{1}{10^{n}}$

47. $\displaystyle\sum_{n=1}^{\infty}(-1)^{n+1}\frac{(0.01)^{n}}{n}$ As you will see in Section 11.7, the sum is $\ln(1.01)$.

48. $\displaystyle\frac{1}{1+t}=\sum_{n=0}^{\infty}(-1)^{n}t^{n},\quad 0<t<1$

T Approximate the sums in Exercises 49 and 50 with an error of magnitude less than 5×10^{-6}.

49. $\displaystyle\sum_{n=0}^{\infty}(-1)^{n}\frac{1}{(2n)!}$ As you will see in Section 11.9, the sum is $\cos 1$, the cosine of 1 radian.

50. $\displaystyle\sum_{n=0}^{\infty}(-1)^{n}\frac{1}{n!}$ As you will see in Section 11.9, the sum is e^{-1}.

Theory and Examples

51. a. The series

$$\frac{1}{3}-\frac{1}{2}+\frac{1}{9}-\frac{1}{4}+\frac{1}{27}-\frac{1}{8}+\cdots+\frac{1}{3^{n}}-\frac{1}{2^{n}}+\cdots$$

does not meet one of the conditions of Theorem 14. Which one?

b. Find the sum of the series in part (a).

T **52.** The limit L of an alternating series that satisfies the conditions of Theorem 14 lies between the values of any two consecutive partial sums. This suggests using the average

$$\frac{s_{n}+s_{n+1}}{2}=s_{n}+\frac{1}{2}(-1)^{n+2}a_{n+1}$$

to estimate L. Compute

$$s_{20}+\frac{1}{2}\cdot\frac{1}{21}$$

as an approximation to the sum of the alternating harmonic series. The exact sum is $\ln 2=0.6931\ldots$.

53. The sign of the remainder of an alternating series that satisfies the conditions of Theorem 14 Prove the assertion in Theorem 15 that whenever an alternating series satisfying the conditions of Theorem 14 is approximated with one of its partial sums, then the remainder (sum of the unused terms) has the same sign as the first unused term. (*Hint:* Group the remainder's terms in consecutive pairs.)

54. Show that the sum of the first $2n$ terms of the series

$$1-\frac{1}{2}+\frac{1}{2}-\frac{1}{3}+\frac{1}{3}-\frac{1}{4}+\frac{1}{4}-\frac{1}{5}+\frac{1}{5}-\frac{1}{6}+\cdots$$

is the same as the sum of the first n terms of the series

$$\frac{1}{1\cdot2}+\frac{1}{2\cdot3}+\frac{1}{3\cdot4}+\frac{1}{4\cdot5}+\frac{1}{5\cdot6}+\cdots.$$

Do these series converge? What is the sum of the first $2n+1$ terms of the first series? If the series converge, what is their sum?

55. Show that if $\sum_{n=1}^{\infty}a_{n}$ diverges, then $\sum_{n=1}^{\infty}|a_{n}|$ diverges.

56. Show that if $\sum_{n=1}^{\infty}a_{n}$ converges absolutely, then

$$\left|\sum_{n=1}^{\infty}a_{n}\right|\leq\sum_{n=1}^{\infty}|a_{n}|.$$

57. Show that if $\sum_{n=1}^{\infty}a_{n}$ and $\sum_{n=1}^{\infty}b_{n}$ both converge absolutely, then so does

a. $\displaystyle\sum_{n=1}^{\infty}(a_{n}+b_{n})$ **b.** $\displaystyle\sum_{n=1}^{\infty}(a_{n}-b_{n})$

c. $\displaystyle\sum_{n=1}^{\infty}ka_{n}$ (k any number)

58. Show by example that $\sum_{n=1}^{\infty}a_{n}b_{n}$ may diverge even if $\sum_{n=1}^{\infty}a_{n}$ and $\sum_{n=1}^{\infty}b_{n}$ both converge.

T **59.** In Example 6, suppose the goal is to arrange the terms to get a new series that converges to $-1/2$. Start the new arrangement with the first negative term, which is $-1/2$. Whenever you have a sum that is less than or equal to $-1/2$, start introducing positive terms, taken in order, until the new total is greater than $-1/2$. Then add negative terms until the total is less than or equal to $-1/2$ again. Continue this process until your partial sums have been above the target at least three times and finish at or below it. If s_n is the sum of the first n terms of your new series, plot the points (n, s_n) to illustrate how the sums are behaving.

60. Outline of the proof of the Rearrangement Theorem (Theorem 17)

a. Let ϵ be a positive real number, let $L = \sum_{n=1}^{\infty} a_n$, and let $s_k = \sum_{n=1}^{k} a_n$. Show that for some index N_1 and for some index $N_2 \geq N_1$,

$$\sum_{n=N_1}^{\infty} |a_n| < \frac{\epsilon}{2} \quad \text{and} \quad |s_{N_2} - L| < \frac{\epsilon}{2}.$$

Since all the terms $a_1, a_2, \ldots, a_{N_2}$ appear somewhere in the sequence $\{b_n\}$, there is an index $N_3 \geq N_2$ such that if $n \geq N_3$, then $\left(\sum_{k=1}^{n} b_k\right) - s_{N_2}$ is at most a sum of terms a_m with $m \geq N_1$. Therefore, if $n \geq N_3$,

$$\left| \sum_{k=1}^{n} b_k - L \right| \leq \left| \sum_{k=1}^{n} b_k - s_{N_2} \right| + |s_{N_2} - L|$$

$$\leq \sum_{k=N_1}^{\infty} |a_k| + |s_{N_2} - L| < \epsilon.$$

b. The argument in part (a) shows that if $\sum_{n=1}^{\infty} a_n$ converges absolutely then $\sum_{n=1}^{\infty} b_n$ converges and $\sum_{n=1}^{\infty} b_n = \sum_{n=1}^{\infty} a_n$. Now show that because $\sum_{n=1}^{\infty} |a_n|$ converges, $\sum_{n=1}^{\infty} |b_n|$ converges to $\sum_{n=1}^{\infty} |a_n|$.

61. Unzipping absolutely convergent series

a. Show that if $\sum_{n=1}^{\infty} |a_n|$ converges and

$$b_n = \begin{cases} a_n, & \text{if } a_n \geq 0 \\ 0, & \text{if } a_n < 0, \end{cases}$$

then $\sum_{n=1}^{\infty} b_n$ converges.

b. Use the results in part (a) to show likewise that if $\sum_{n=1}^{\infty} |a_n|$ converges and

$$c_n = \begin{cases} 0, & \text{if } a_n \geq 0 \\ a_n, & \text{if } a_n < 0, \end{cases}$$

then $\sum_{n=1}^{\infty} c_n$ converges.

In other words, if a series converges absolutely, its positive terms form a convergent series, and so do its negative terms. Furthermore,

$$\sum_{n=1}^{\infty} a_n = \sum_{n=1}^{\infty} b_n + \sum_{n=1}^{\infty} c_n$$

because $b_n = (a_n + |a_n|)/2$ and $c_n = (a_n - |a_n|)/2$.

62. What is wrong here?:

Multiply both sides of the alternating harmonic series

$$S = 1 - \frac{1}{2} + \frac{1}{3} - \frac{1}{4} + \frac{1}{5} - \frac{1}{6} +$$
$$\frac{1}{7} - \frac{1}{8} + \frac{1}{9} - \frac{1}{10} + \frac{1}{11} - \frac{1}{12} + \cdots$$

by 2 to get

$$2S = 2 - 1 +$$

$$\frac{2}{3} - \frac{1}{2} + \frac{2}{5} - \frac{1}{3} + \frac{2}{7} - \frac{1}{4} + \frac{2}{9} - \frac{1}{5} + \frac{2}{11} - \frac{1}{6} + \cdots.$$

Collect terms with the same denominator, as the arrows indicate, to arrive at

$$2S = 1 - \frac{1}{2} + \frac{1}{3} - \frac{1}{4} + \frac{1}{5} - \frac{1}{6} + \cdots.$$

The series on the right-hand side of this equation is the series we started with. Therefore, $2S = S$, and dividing by S gives $2 = 1$. (*Source:* "Riemann's Rearrangement Theorem" by Stewart Galanor, *Mathematics Teacher*, Vol. 80, No. 8, 1987, pp. 675–681.)

63. Draw a figure similar to Figure 11.9 to illustrate the convergence of the series in Theorem 14 when $N > 1$.

11.7 Power Series

Now that we can test infinite series for convergence we can study the infinite polynomials mentioned at the beginning of this chapter. We call these polynomials power series because they are defined as infinite series of powers of some variable, in our case x. Like polynomials, power series can be added, subtracted, multiplied, differentiated, and integrated to give new power series.

Power Series and Convergence

We begin with the formal definition.

DEFINITIONS   **Power Series, Center, Coefficients**

A **power series about $x = 0$** is a series of the form

$$\sum_{n=0}^{\infty} c_n x^n = c_0 + c_1 x + c_2 x^2 + \cdots + c_n x^n + \cdots. \qquad (1)$$

A **power series about $x = a$** is a series of the form

$$\sum_{n=0}^{\infty} c_n (x - a)^n = c_0 + c_1(x - a) + c_2(x - a)^2 + \cdots + c_n(x - a)^n + \cdots \qquad (2)$$

in which the **center** a and the **coefficients** $c_0, c_1, c_2, \ldots, c_n, \ldots$ are constants.

Equation (1) is the special case obtained by taking $a = 0$ in Equation (2).

EXAMPLE 1   A Geometric Series

Taking all the coefficients to be 1 in Equation (1) gives the geometric power series

$$\sum_{n=0}^{\infty} x^n = 1 + x + x^2 + \cdots + x^n + \cdots.$$

This is the geometric series with first term 1 and ratio x. It converges to $1/(1 - x)$ for $|x| < 1$. We express this fact by writing

$$\frac{1}{1 - x} = 1 + x + x^2 + \cdots + x^n + \cdots, \qquad -1 < x < 1. \qquad (3)$$

■

Up to now, we have used Equation (3) as a formula for the sum of the series on the right. We now change the focus: We think of the partial sums of the series on the right as polynomials $P_n(x)$ that approximate the function on the left. For values of x near zero, we need take only a few terms of the series to get a good approximation. As we move toward $x = 1$, or -1, we must take more terms. Figure 11.10 shows the graphs of $f(x) = 1/(1 - x)$, and the approximating polynomials $y_n = P_n(x)$ for $n = 0, 1, 2,$ and 8. The function $f(x) = 1/(1 - x)$ is not continuous on intervals containing $x = 1$, where it has a vertical asymptote. The approximations do not apply when $x \geq 1$.

EXAMPLE 2   A Geometric Series

The power series

$$1 - \frac{1}{2}(x - 2) + \frac{1}{4}(x - 2)^2 + \cdots + \left(-\frac{1}{2}\right)^n (x - 2)^n + \cdots \qquad (4)$$

matches Equation (2) with $a = 2, c_0 = 1, c_1 = -1/2, c_2 = 1/4, \ldots, c_n = (-1/2)^n$. This is a geometric series with first term 1 and ratio $r = -\dfrac{x - 2}{2}$. The series converges for

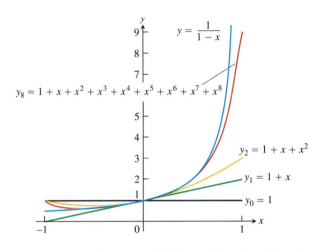

FIGURE 11.10 The graphs of $f(x) = 1/(1 - x)$ and four of its polynomial approximations (Example 1).

$\left| \dfrac{x - 2}{2} \right| < 1$ or $0 < x < 4$. The sum is

$$\frac{1}{1 - r} = \frac{1}{1 + \dfrac{x - 2}{2}} = \frac{2}{x},$$

so

$$\frac{2}{x} = 1 - \frac{(x - 2)}{2} + \frac{(x - 2)^2}{4} - \cdots + \left(-\frac{1}{2}\right)^n (x - 2)^n + \cdots, \qquad 0 < x < 4.$$

Series (4) generates useful polynomial approximations of $f(x) = 2/x$ for values of x near 2:

$$P_0(x) = 1$$

$$P_1(x) = 1 - \frac{1}{2}(x - 2) = 2 - \frac{x}{2}$$

$$P_2(x) = 1 - \frac{1}{2}(x - 2) + \frac{1}{4}(x - 2)^2 = 3 - \frac{3x}{2} + \frac{x^2}{4},$$

and so on (Figure 11.11). ∎

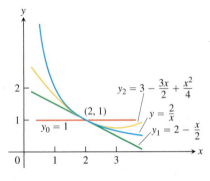

FIGURE 11.11 The graphs of $f(x) = 2/x$ and its first three polynomial approximations (Example 2).

EXAMPLE 3   Testing for Convergence Using the Ratio Test

For what values of x do the following power series converge?

(a) $\displaystyle\sum_{n=1}^{\infty} (-1)^{n-1} \frac{x^n}{n} = x - \frac{x^2}{2} + \frac{x^3}{3} - \cdots$

(b) $\displaystyle\sum_{n=1}^{\infty} (-1)^{n-1} \frac{x^{2n-1}}{2n - 1} = x - \frac{x^3}{3} + \frac{x^5}{5} - \cdots$

(c) $\displaystyle\sum_{n=0}^{\infty} \frac{x^n}{n!} = 1 + x + \frac{x^2}{2!} + \frac{x^3}{3!} + \cdots$

(d) $\displaystyle\sum_{n=0}^{\infty} n! x^n = 1 + x + 2! x^2 + 3! x^3 + \cdots$

Solution   Apply the Ratio Test to the series $\sum |u_n|$, where u_n is the nth term of the series in question.

(a) $\left| \dfrac{u_{n+1}}{u_n} \right| = \dfrac{n}{n+1} |x| \to |x|.$

The series converges absolutely for $|x| < 1$. It diverges if $|x| > 1$ because the nth term does not converge to zero. At $x = 1$, we get the alternating harmonic series $1 - 1/2 + 1/3 - 1/4 + \cdots$, which converges. At $x = -1$ we get $-1 - 1/2 - 1/3 - 1/4 - \cdots$, the negative of the harmonic series; it diverges. Series (a) converges for $-1 < x \le 1$ and diverges elsewhere.

(b) $\left| \dfrac{u_{n+1}}{u_n} \right| = \dfrac{2n-1}{2n+1} x^2 \to x^2.$

The series converges absolutely for $x^2 < 1$. It diverges for $x^2 > 1$ because the nth term does not converge to zero. At $x = 1$ the series becomes $1 - 1/3 + 1/5 - 1/7 + \cdots$, which converges by the Alternating Series Theorem. It also converges at $x = -1$ because it is again an alternating series that satisfies the conditions for convergence. The value at $x = -1$ is the negative of the value at $x = 1$. Series (b) converges for $-1 \le x \le 1$ and diverges elsewhere.

(c) $\left| \dfrac{u_{n+1}}{u_n} \right| = \left| \dfrac{x^{n+1}}{(n+1)!} \cdot \dfrac{n!}{x^n} \right| = \dfrac{|x|}{n+1} \to 0 \text{ for every } x.$

The series converges absolutely for all x.

(d) $\left| \dfrac{u_{n+1}}{u_n} \right| = \left| \dfrac{(n+1)! x^{n+1}}{n! x^n} \right| = (n+1)|x| \to \infty \text{ unless } x = 0.$

The series diverges for all values of x except $x = 0$.

Example 3 illustrates how we usually test a power series for convergence, and the possible results.

THEOREM 18   The Convergence Theorem for Power Series

If the power series $\displaystyle\sum_{n=0}^{\infty} a_n x^n = a_0 + a_1 x + a_2 x^2 + \cdots$ converges for $x = c \ne 0$, then it converges absolutely for all x with $|x| < |c|$. If the series diverges for $x = d$, then it diverges for all x with $|x| > |d|$.

Proof   Suppose the series $\sum_{n=0}^{\infty} a_n c^n$ converges. Then $\lim_{n \to \infty} a_n c^n = 0$. Hence, there is an integer N such that $|a_n c^n| < 1$ for all $n \geq N$. That is,

$$|a_n| < \frac{1}{|c|^n} \qquad \text{for } n \geq N. \tag{5}$$

Now take any x such that $|x| < |c|$ and consider

$$|a_0| + |a_1 x| + \cdots + |a_{N-1} x^{N-1}| + |a_N x^N| + |a_{N+1} x^{N+1}| + \cdots .$$

There are only a finite number of terms prior to $|a_N x^N|$, and their sum is finite. Starting with $|a_N x^N|$ and beyond, the terms are less than

$$\left|\frac{x}{c}\right|^N + \left|\frac{x}{c}\right|^{N+1} + \left|\frac{x}{c}\right|^{N+2} + \cdots \tag{6}$$

because of Inequality (5). But Series (6) is a geometric series with ratio $r = |x/c|$, which is less than 1, since $|x| < |c|$. Hence Series (6) converges, so the original series converges absolutely. This proves the first half of the theorem.

The second half of the theorem follows from the first. If the series diverges at $x = d$ and converges at a value x_0 with $|x_0| > |d|$, we may take $c = x_0$ in the first half of the theorem and conclude that the series converges absolutely at d. But the series cannot converge absolutely and diverge at one and the same time. Hence, if it diverges at d, it diverges for all x with $|x| > |d|$.   ∎

To simplify the notation, Theorem 18 deals with the convergence of series of the form $\sum a_n x^n$. For series of the form $\sum a_n (x - a)^n$ we can replace $x - a$ by x' and apply the results to the series $\sum a_n (x')^n$.

The Radius of Convergence of a Power Series

The theorem we have just proved and the examples we have studied lead to the conclusion that a power series $\sum c_n (x - a)^n$ behaves in one of three possible ways. It might converge only at $x = a$, or converge everywhere, or converge on some interval of radius R centered at $x = a$. We prove this as a Corollary to Theorem 18.

COROLLARY TO THEOREM 18

The convergence of the series $\sum c_n (x - a)^n$ is described by one of the following three possibilities:

1.   There is a positive number R such that the series diverges for x with $|x - a| > R$ but converges absolutely for x with $|x - a| < R$. The series may or may not converge at either of the endpoints $x = a - R$ and $x = a + R$.

2.   The series converges absolutely for every x ($R = \infty$).

3.   The series converges at $x = a$ and diverges elsewhere ($R = 0$).

Proof   We assume first that $a = 0$, so that the power series is centered at 0. If the series converges everywhere we are in Case 2. If it converges only at $x = 0$ we are in Case 3. Otherwise there is a nonzero number d such that $\sum c_n d^n$ diverges. The set S of values of x for which the series $\sum c_n x^n$ converges is nonempty because it contains 0 and a positive number p as well. By Theorem 18, the series diverges for all x with $|x| > |d|$, so $|x| \leq |d|$ for all $x \in S$, and S is a bounded set. By the Completeness Property of the real numbers (see Appendix 4) a nonempty, bounded set has a least upper bound R. (The least upper bound is the smallest number with the property that the elements $x \in S$ satisfy $x \leq R$.) If $|x| > R \geq p$, then $x \notin S$ so the series $\sum c_n x^n$ diverges. If $|x| < R$, then $|x|$ is not an upper bound for S (because it's smaller than the least upper bound) so there is a number $b \in S$ such that $b > |x|$. Since $b \in S$, the series $\sum c_n b^n$ converges and therefore the series $\sum c_n |x|^n$ converges by Theorem 18. This proves the Corollary for power series centered at $a = 0$.

For a power series centered at $a \neq 0$, we set $x' = (x - a)$ and repeat the argument with x'. Since $x' = 0$ when $x = a$, a radius R interval of convergence for $\sum c_n (x')^n$ centered at $x' = 0$ is the same as a radius R interval of convergence for $\sum c_n (x - a)^n$ centered at $x = a$. This establishes the Corollary for the general case.   ∎

R is called the **radius of convergence** of the power series and the interval of radius R centered at $x = a$ is called the **interval of convergence**. The interval of convergence may be open, closed, or half-open, depending on the particular series. At points x with $|x - a| < R$, the series converges absolutely. If the series converges for all values of x, we say its radius of convergence is infinite. If it converges only at $x = a$, we say its radius of convergence is zero.

How to Test a Power Series for Convergence

1. *Use the Ratio Test (or nth-Root Test) to find the interval where the series converges absolutely.* Ordinarily, this is an open interval

    $$|x - a| < R \qquad \text{or} \qquad a - R < x < a + R.$$

2. *If the interval of absolute convergence is finite, test for convergence or divergence at each endpoint,* as in Examples 3a and b. Use a Comparison Test, the Integral Test, or the Alternating Series Test.

3. *If the interval of absolute convergence is $a - R < x < a + R$, the series diverges for $|x - a| > R$* (it does not even converge conditionally), because the nth term does not approach zero for those values of x.

Term-by-Term Differentiation

A theorem from advanced calculus says that a power series can be differentiated term by term at each interior point of its interval of convergence.

THEOREM 19   The Term-by-Term Differentiation Theorem

If $\sum c_n(x - a)^n$ converges for $a - R < x < a + R$ for some $R > 0$, it defines a function f:

$$f(x) = \sum_{n=0}^{\infty} c_n(x - a)^n, \qquad a - R < x < a + R.$$

Such a function f has derivatives of all orders inside the interval of convergence. We can obtain the derivatives by differentiating the original series term by term:

$$f'(x) = \sum_{n=1}^{\infty} nc_n(x - a)^{n-1}$$

$$f''(x) = \sum_{n=2}^{\infty} n(n - 1)c_n(x - a)^{n-2},$$

and so on. Each of these derived series converges at every interior point of the interval of convergence of the original series.

EXAMPLE 4   Applying Term-by-Term Differentiation

Find series for $f'(x)$ and $f''(x)$ if

$$f(x) = \frac{1}{1 - x} = 1 + x + x^2 + x^3 + x^4 + \cdots + x^n + \cdots$$

$$= \sum_{n=0}^{\infty} x^n, \qquad -1 < x < 1$$

Solution

$$f'(x) = \frac{1}{(1 - x)^2} = 1 + 2x + 3x^2 + 4x^3 + \cdots + nx^{n-1} + \cdots$$

$$= \sum_{n=1}^{\infty} nx^{n-1}, \qquad -1 < x < 1$$

$$f''(x) = \frac{2}{(1 - x)^3} = 2 + 6x + 12x^2 + \cdots + n(n - 1)x^{n-2} + \cdots$$

$$= \sum_{n=2}^{\infty} n(n - 1)x^{n-2}, \qquad -1 < x < 1 \qquad \blacksquare$$

CAUTION   Term-by-term differentiation might not work for other kinds of series. For example, the trigonometric series

$$\sum_{n=1}^{\infty} \frac{\sin (n!x)}{n^2}$$

converges for all x. But if we differentiate term by term we get the series

$$\sum_{n=1}^{\infty} \frac{n!\cos (n!x)}{n^2},$$

which diverges for all x. This is not a power series, since it is not a sum of positive integer powers of x.

Term-by-Term Integration

Another advanced calculus theorem states that a power series can be integrated term by term throughout its interval of convergence.

THEOREM 20   **The Term-by-Term Integration Theorem**

Suppose that

$$f(x) = \sum_{n=0}^{\infty} c_n(x - a)^n$$

converges for $a - R < x < a + R$ $(R > 0)$. Then

$$\sum_{n=0}^{\infty} c_n \frac{(x - a)^{n+1}}{n + 1}$$

converges for $a - R < x < a + R$ and

$$\int f(x)\, dx = \sum_{n=0}^{\infty} c_n \frac{(x - a)^{n+1}}{n + 1} + C$$

for $a - R < x < a + R$.

EXAMPLE 5   A Series for $\tan^{-1} x$, $-1 \le x \le 1$

Identify the function

$$f(x) = x - \frac{x^3}{3} + \frac{x^5}{5} - \cdots, \qquad -1 \le x \le 1.$$

Solution   We differentiate the original series term by term and get

$$f'(x) = 1 - x^2 + x^4 - x^6 + \cdots, \qquad -1 < x < 1.$$

This is a geometric series with first term 1 and ratio $-x^2$, so

$$f'(x) = \frac{1}{1 - (-x^2)} = \frac{1}{1 + x^2}.$$

We can now integrate $f'(x) = 1/(1 + x^2)$ to get

$$\int f'(x)\, dx = \int \frac{dx}{1 + x^2} = \tan^{-1} x + C.$$

The series for $f(x)$ is zero when $x = 0$, so $C = 0$. Hence

$$f(x) = x - \frac{x^3}{3} + \frac{x^5}{5} - \frac{x^7}{7} + \cdots = \tan^{-1} x, \qquad -1 < x < 1. \tag{7}$$

In Section 11.10, we will see that the series also converges to $\tan^{-1} x$ at $x = \pm 1$.   ■

Notice that the original series in Example 5 converges at both endpoints of the original interval of convergence, but Theorem 20 can guarantee the convergence of the differentiated series only inside the interval.

EXAMPLE 6   A Series for $\ln (1 + x)$, $-1 < x \leq 1$

The series

$$\frac{1}{1 + t} = 1 - t + t^2 - t^3 + \cdots$$

converges on the open interval $-1 < t < 1$. Therefore,

$$\ln (1 + x) = \int_0^x \frac{1}{1 + t}\, dt = t - \frac{t^2}{2} + \frac{t^3}{3} - \frac{t^4}{4} + \cdots \bigg]_0^x \qquad \text{Theorem 20}$$

$$= x - \frac{x^2}{2} + \frac{x^3}{3} - \frac{x^4}{4} + \cdots, \qquad -1 < x < 1.$$

It can also be shown that the series converges at $x = 1$ to the number $\ln 2$, but that was not guaranteed by the theorem.   ∎

USING TECHNOLOGY   Study of Series

Series are in many ways analogous to integrals. Just as the number of functions with explicit antiderivatives in terms of elementary functions is small compared to the number of integrable functions, the number of power series in x that agree with explicit elementary functions on x-intervals is small compared to the number of power series that converge on some x-interval. Graphing utilities can aid in the study of such series in much the same way that numerical integration aids in the study of definite integrals. The ability to study power series at particular values of x is built into most Computer Algebra Systems.

If a series converges rapidly enough, CAS exploration might give us an idea of the sum. For instance, in calculating the early partial sums of the series $\sum_{k=1}^{\infty} [1/(2^{k-1})]$ (Section 11.4, Example 2b), Maple returns $S_n = 1.6066\ 95152$ for $31 \leq n \leq 200$. This suggests that the sum of the series is 1.6066 95152 to 10 digits. Indeed,

$$\sum_{k=201}^{\infty} \frac{1}{2^k - 1} = \sum_{k=201}^{\infty} \frac{1}{2^{k-1}(2 - (1/2^{k-1}))} < \sum_{k=201}^{\infty} \frac{1}{2^{k-1}} = \frac{1}{2^{199}} < 1.25 \times 10^{-60}.$$

The remainder after 200 terms is negligible.

However, CAS and calculator exploration cannot do much for us if the series converges or diverges very slowly, and indeed can be downright misleading. For example, try calculating the partial sums of the series $\sum_{k=1}^{\infty} [1/(10^{10}k)]$. The terms are tiny in comparison to the numbers we normally work with and the partial sums, even for hundreds of terms, are miniscule. We might well be fooled into thinking that the series converges. In fact, it diverges, as we can see by writing it as $(1/10^{10}) \sum_{k=1}^{\infty} (1/k)$, a constant times the harmonic series.

We will know better how to interpret numerical results after studying error estimates in Section 11.9.

Multiplication of Power Series

Another theorem from advanced calculus states that absolutely converging power series can be multiplied the way we multiply polynomials. We omit the proof.

THEOREM 21   The Series Multiplication Theorem for Power Series

If $A(x) = \sum_{n=0}^{\infty} a_n x^n$ and $B(x) = \sum_{n=0}^{\infty} b_n x^n$ converge absolutely for $|x| < R$, and

$$c_n = a_0 b_n + a_1 b_{n-1} + a_2 b_{n-2} + \cdots + a_{n-1} b_1 + a_n b_0 = \sum_{k=0}^{n} a_k b_{n-k},$$

then $\sum_{n=0}^{\infty} c_n x^n$ converges absolutely to $A(x)B(x)$ for $|x| < R$:

$$\left(\sum_{n=0}^{\infty} a_n x^n \right) \cdot \left(\sum_{n=0}^{\infty} b_n x^n \right) = \sum_{n=0}^{\infty} c_n x^n.$$

EXAMPLE 7   Multiply the geometric series

$$\sum_{n=0}^{\infty} x^n = 1 + x + x^2 + \cdots + x^n + \cdots = \frac{1}{1-x}, \qquad \text{for } |x| < 1,$$

by itself to get a power series for $1/(1-x)^2$, for $|x| < 1$.

Solution   Let

$$A(x) = \sum_{n=0}^{\infty} a_n x^n = 1 + x + x^2 + \cdots + x^n + \cdots = 1/(1-x)$$

$$B(x) = \sum_{n=0}^{\infty} b_n x^n = 1 + x + x^2 + \cdots + x^n + \cdots = 1/(1-x)$$

and

$$c_n = \underbrace{a_0 b_n + a_1 b_{n-1} + \cdots + a_k b_{n-k} + \cdots + a_n b_0}_{n+1 \text{ terms}}$$

$$= \underbrace{1 + 1 + \cdots + 1}_{n+1 \text{ ones}} = n + 1.$$

Then, by the Series Multiplication Theorem,

$$A(x) \cdot B(x) = \sum_{n=0}^{\infty} c_n x^n = \sum_{n=0}^{\infty} (n+1) x^n$$

$$= 1 + 2x + 3x^2 + 4x^3 + \cdots + (n+1)x^n + \cdots$$

is the series for $1/(1-x)^2$. The series all converge absolutely for $|x| < 1$.

Notice that Example 4 gives the same answer because

$$\frac{d}{dx} \left(\frac{1}{1-x} \right) = \frac{1}{(1-x)^2}.$$

EXERCISES 11.7

Intervals of Convergence

In Exercises 1–32, **(a)** find the series' radius and interval of convergence. For what values of x does the series converge **(b)** absolutely, **(c)** conditionally?

1. $\sum_{n=0}^{\infty} x^n$

2. $\sum_{n=0}^{\infty} (x + 5)^n$

3. $\sum_{n=0}^{\infty} (-1)^n (4x + 1)^n$

4. $\sum_{n=1}^{\infty} \frac{(3x - 2)^n}{n}$

5. $\sum_{n=0}^{\infty} \frac{(x - 2)^n}{10^n}$

6. $\sum_{n=0}^{\infty} (2x)^n$

7. $\sum_{n=0}^{\infty} \frac{nx^n}{n + 2}$

8. $\sum_{n=1}^{\infty} \frac{(-1)^n (x + 2)^n}{n}$

9. $\sum_{n=1}^{\infty} \frac{x^n}{n\sqrt{n}\, 3^n}$

10. $\sum_{n=1}^{\infty} \frac{(x - 1)^n}{\sqrt{n}}$

11. $\sum_{n=0}^{\infty} \frac{(-1)^n x^n}{n!}$

12. $\sum_{n=0}^{\infty} \frac{3^n x^n}{n!}$

13. $\sum_{n=0}^{\infty} \frac{x^{2n+1}}{n!}$

14. $\sum_{n=0}^{\infty} \frac{(2x + 3)^{2n+1}}{n!}$

15. $\sum_{n=0}^{\infty} \frac{x^n}{\sqrt{n^2 + 3}}$

16. $\sum_{n=0}^{\infty} \frac{(-1)^n x^n}{\sqrt{n^2 + 3}}$

17. $\sum_{n=0}^{\infty} \frac{n(x + 3)^n}{5^n}$

18. $\sum_{n=0}^{\infty} \frac{nx^n}{4^n(n^2 + 1)}$

19. $\sum_{n=0}^{\infty} \frac{\sqrt{n}\, x^n}{3^n}$

20. $\sum_{n=1}^{\infty} \sqrt[n]{n}\,(2x + 5)^n$

21. $\sum_{n=1}^{\infty} \left(1 + \frac{1}{n}\right)^n x^n$

22. $\sum_{n=1}^{\infty} (\ln n) x^n$

23. $\sum_{n=1}^{\infty} n^n x^n$

24. $\sum_{n=0}^{\infty} n!(x - 4)^n$

25. $\sum_{n=1}^{\infty} \frac{(-1)^{n+1}(x + 2)^n}{n 2^n}$

26. $\sum_{n=0}^{\infty} (-2)^n (n + 1)(x - 1)^n$

27. $\sum_{n=2}^{\infty} \frac{x^n}{n(\ln n)^2}$
Get the information you need about $\sum 1/(n(\ln n)^2)$ from Section 11.3, Exercise 39.

28. $\sum_{n=2}^{\infty} \frac{x^n}{n \ln n}$
Get the information you need about $\sum 1/(n \ln n)$ from Section 11.3, Exercise 38.

29. $\sum_{n=1}^{\infty} \frac{(4x - 5)^{2n+1}}{n^{3/2}}$

30. $\sum_{n=1}^{\infty} \frac{(3x + 1)^{n+1}}{2n + 2}$

31. $\sum_{n=1}^{\infty} \frac{(x + \pi)^n}{\sqrt{n}}$

32. $\sum_{n=0}^{\infty} \frac{(x - \sqrt{2})^{2n+1}}{2^n}$

In Exercises 33–38, find the series' interval of convergence and, within this interval, the sum of the series as a function of x.

33. $\sum_{n=0}^{\infty} \frac{(x - 1)^{2n}}{4n}$

34. $\sum_{n=0}^{\infty} \frac{(x + 1)^{2n}}{9^n}$

35. $\sum_{n=0}^{\infty} \left(\frac{\sqrt{x}}{2} - 1\right)^n$

36. $\sum_{n=0}^{\infty} (\ln x)^n$

37. $\sum_{n=0}^{\infty} \left(\frac{x^2 + 1}{3}\right)^n$

38. $\sum_{n=0}^{\infty} \left(\frac{x^2 - 1}{2}\right)^n$

Theory and Examples

39. For what values of x does the series

$$1 - \frac{1}{2}(x - 3) + \frac{1}{4}(x - 3)^2 + \cdots + \left(-\frac{1}{2}\right)^n (x - 3)^n + \cdots$$

converge? What is its sum? What series do you get if you differentiate the given series term by term? For what values of x does the new series converge? What is its sum?

40. If you integrate the series in Exercise 39 term by term, what new series do you get? For what values of x does the new series converge, and what is another name for its sum?

41. The series

$$\sin x = x - \frac{x^3}{3!} + \frac{x^5}{5!} - \frac{x^7}{7!} + \frac{x^9}{9!} - \frac{x^{11}}{11!} + \cdots$$

converges to $\sin x$ for all x.

a. Find the first six terms of a series for $\cos x$. For what values of x should the series converge?

b. By replacing x by $2x$ in the series for $\sin x$, find a series that converges to $\sin 2x$ for all x.

c. Using the result in part (a) and series multiplication, calculate the first six terms of a series for $2 \sin x \cos x$. Compare your answer with the answer in part (b).

42. The series

$$e^x = 1 + x + \frac{x^2}{2!} + \frac{x^3}{3!} + \frac{x^4}{4!} + \frac{x^5}{5!} + \cdots$$

converges to e^x for all x.

a. Find a series for $(d/dx)e^x$. Do you get the series for e^x? Explain your answer.

b. Find a series for $\int e^x \, dx$. Do you get the series for e^x? Explain your answer.

c. Replace x by $-x$ in the series for e^x to find a series that converges to e^{-x} for all x. Then multiply the series for e^x and e^{-x} to find the first six terms of a series for $e^{-x} \cdot e^x$.

43. The series

$$\tan x = x + \frac{x^3}{3} + \frac{2x^5}{15} + \frac{17x^7}{315} + \frac{62x^9}{2835} + \cdots$$

converges to $\tan x$ for $-\pi/2 < x < \pi/2$.

a. Find the first five terms of the series for $\ln|\sec x|$. For what values of x should the series converge?

b. Find the first five terms of the series for $\sec^2 x$. For what values of x should this series converge?

c. Check your result in part (b) by squaring the series given for $\sec x$ in Exercise 44.

44. The series

$$\sec x = 1 + \frac{x^2}{2} + \frac{5}{24}x^4 + \frac{61}{720}x^6 + \frac{277}{8064}x^8 + \cdots$$

converges to $\sec x$ for $-\pi/2 < x < \pi/2$.

a. Find the first five terms of a power series for the function $\ln|\sec x + \tan x|$. For what values of x should the series converge?

b. Find the first four terms of a series for $\sec x \tan x$. For what values of x should the series converge?

c. Check your result in part (b) by multiplying the series for $\sec x$ by the series given for $\tan x$ in Exercise 43.

45. Uniqueness of convergent power series

a. Show that if two power series $\sum_{n=0}^{\infty} a_n x^n$ and $\sum_{n=0}^{\infty} b_n x^n$ are convergent and equal for all values of x in an open interval $(-c, c)$, then $a_n = b_n$ for every n. (*Hint:* Let $f(x) = \sum_{n=0}^{\infty} a_n x^n = \sum_{n=0}^{\infty} b_n x^n$. Differentiate term by term to show that a_n and b_n both equal $f^{(n)}(0)/(n!)$.)

b. Show that if $\sum_{n=0}^{\infty} a_n x^n = 0$ for all x in an open interval $(-c, c)$, then $a_n = 0$ for every n.

46. The sum of the series $\sum_{n=0}^{\infty} (n^2/2^n)$ To find the sum of this series, express $1/(1 - x)$ as a geometric series, differentiate both sides of the resulting equation with respect to x, multiply both sides of the result by x, differentiate again, multiply by x again, and set x equal to $1/2$. What do you get? (*Source*: David E. Dobbs' letter to the editor, *Illinois Mathematics Teacher*, Vol. 33, Issue 4, 1982, p. 27.)

47. Convergence at endpoints Show by examples that the convergence of a power series at an endpoint of its interval of convergence may be either conditional or absolute.

48. Make up a power series whose interval of convergence is

a. $(-3, 3)$ **b.** $(-2, 0)$ **c.** $(1, 5)$.

11.8 Taylor and Maclaurin Series

This section shows how functions that are infinitely differentiable generate power series called Taylor series. In many cases, these series can provide useful polynomial approximations of the generating functions.

Series Representations

We know from Theorem 19 that within its interval of convergence the sum of a power series is a continuous function with derivatives of all orders. But what about the other way around? If a function $f(x)$ has derivatives of all orders on an interval I, can it be expressed as a power series on I? And if it can, what will its coefficients be?

We can answer the last question readily if we assume that $f(x)$ is the sum of a power series

$$f(x) = \sum_{n=0}^{\infty} a_n(x - a)^n$$

$$= a_0 + a_1(x - a) + a_2(x - a)^2 + \cdots + a_n(x - a)^n + \cdots$$

with a positive radius of convergence. By repeated term-by-term differentiation within the interval of convergence I we obtain

$$f'(x) = a_1 + 2a_2(x - a) + 3a_3(x - a)^2 + \cdots + na_n(x - a)^{n-1} + \cdots$$

$$f''(x) = 1 \cdot 2a_2 + 2 \cdot 3a_3(x - a) + 3 \cdot 4a_4(x - a)^2 + \cdots$$

$$f'''(x) = 1 \cdot 2 \cdot 3a_3 + 2 \cdot 3 \cdot 4a_4(x - a) + 3 \cdot 4 \cdot 5a_5(x - a)^2 + \cdots,$$

with the nth derivative, for all n, being

$$f^{(n)}(x) = n!a_n + \text{a sum of terms with } (x - a) \text{ as a factor}.$$

Since these equations all hold at $x = a$, we have

$$f'(a) = a_1,$$
$$f''(a) = 1 \cdot 2a_2,$$
$$f'''(a) = 1 \cdot 2 \cdot 3a_3,$$

and, in general,

$$f^{(n)}(a) = n!a_n.$$

These formulas reveal a pattern in the coefficients of any power series $\sum_{n=0}^{\infty} a_n(x - a)^n$ that converges to the values of f on I ("represents f on I"). If there *is* such a series (still an open question), then there is only one such series and its nth coefficient is

$$a_n = \frac{f^{(n)}(a)}{n!}.$$

If f has a series representation, then the series must be

$$f(x) = f(a) + f'(a)(x - a) + \frac{f''(a)}{2!}(x - a)^2$$

$$+ \cdots + \frac{f^{(n)}(a)}{n!}(x - a)^n + \cdots. \tag{1}$$

But if we start with an arbitrary function f that is infinitely differentiable on an interval I centered at $x = a$ and use it to generate the series in Equation (1), will the series then converge to $f(x)$ at each x in the interior of I? The answer is maybe—for some functions it will but for other functions it will not, as we will see.

Taylor and Maclaurin Series

DEFINITIONS   Taylor Series, Maclaurin Series

Let f be a function with derivatives of all orders throughout some interval containing a as an interior point. Then the **Taylor series generated by f at $x = a$** is

$$\sum_{k=0}^{\infty} \frac{f^{(k)}(a)}{k!}(x - a)^k = f(a) + f'(a)(x - a) + \frac{f''(a)}{2!}(x - a)^2$$

$$+ \cdots + \frac{f^{(n)}(a)}{n!}(x - a)^n + \cdots.$$

The **Maclaurin series generated by f** is

$$\sum_{k=0}^{\infty} \frac{f^{(k)}(0)}{k!}x^k = f(0) + f'(0)x + \frac{f''(0)}{2!}x^2 + \cdots + \frac{f^{(n)}(0)}{n!}x^n + \cdots,$$

the Taylor series generated by f at $x = 0$.

The Maclaurin series generated by f is often just called the Taylor series of f.

EXAMPLE 1   Finding a Taylor Series

Find the Taylor series generated by $f(x) = 1/x$ at $a = 2$. Where, if anywhere, does the series converge to $1/x$?

Solution   We need to find $f(2), f'(2), f''(2), \ldots$. Taking derivatives we get

$$f(x) = x^{-1}, \qquad\qquad f(2) = 2^{-1} = \frac{1}{2},$$

$$f'(x) = -x^{-2}, \qquad\qquad f'(2) = -\frac{1}{2^2},$$

$$f''(x) = 2!x^{-3}, \qquad\qquad \frac{f''(2)}{2!} = 2^{-3} = \frac{1}{2^3},$$

$$f'''(x) = -3!x^{-4}, \qquad\qquad \frac{f'''(2)}{3!} = -\frac{1}{2^4},$$

$$\vdots \qquad\qquad\qquad\qquad \vdots$$

$$f^{(n)}(x) = (-1)^n n! x^{-(n+1)}, \qquad \frac{f^{(n)}(2)}{n!} = \frac{(-1)^n}{2^{n+1}}.$$

The Taylor series is

$$f(2) + f'(2)(x - 2) + \frac{f''(2)}{2!}(x - 2)^2 + \cdots + \frac{f^{(n)}(2)}{n!}(x - 2)^n + \cdots$$

$$= \frac{1}{2} - \frac{(x - 2)}{2^2} + \frac{(x - 2)^2}{2^3} - \cdots + (-1)^n \frac{(x - 2)^n}{2^{n+1}} + \cdots.$$

This is a geometric series with first term $1/2$ and ratio $r = -(x - 2)/2$. It converges absolutely for $|x - 2| < 2$ and its sum is

$$\frac{1/2}{1 + (x - 2)/2} = \frac{1}{2 + (x - 2)} = \frac{1}{x}.$$

In this example the Taylor series generated by $f(x) = 1/x$ at $a = 2$ converges to $1/x$ for $|x - 2| < 2$ or $0 < x < 4$.   ∎

Taylor Polynomials

The linearization of a differentiable function f at a point a is the polynomial of degree one given by

$$P_1(x) = f(a) + f'(a)(x - a).$$

In Section 3.8 we used this linearization to approximate $f(x)$ at values of x near a. If f has derivatives of higher order at a, then it has higher-order polynomial approximations as well, one for each available derivative. These polynomials are called the Taylor polynomials of f.

DEFINITION **Taylor Polynomial of Order *n***

Let f be a function with derivatives of order k for $k = 1, 2, \ldots, N$ in some interval containing a as an interior point. Then for any integer n from 0 through N, the **Taylor polynomial of order *n*** generated by f at $x = a$ is the polynomial

$$P_n(x) = f(a) + f'(a)(x - a) + \frac{f''(a)}{2!}(x - a)^2 + \cdots$$

$$+ \frac{f^{(k)}(a)}{k!}(x - a)^k + \cdots + \frac{f^{(n)}(a)}{n!}(x - a)^n.$$

We speak of a Taylor polynomial of *order n* rather than *degree n* because $f^{(n)}(a)$ may be zero. The first two Taylor polynomials of $f(x) = \cos x$ at $x = 0$, for example, are $P_0(x) = 1$ and $P_1(x) = 1$. The first-order Taylor polynomial has degree zero, not one.

Just as the linearization of f at $x = a$ provides the best linear approximation of f in the neighborhood of a, the higher-order Taylor polynomials provide the best polynomial approximations of their respective degrees. (See Exercise 32.)

EXAMPLE 2 Finding Taylor Polynomials for e^x

Find the Taylor series and the Taylor polynomials generated by $f(x) = e^x$ at $x = 0$.

Solution Since

$$f(x) = e^x, \qquad f'(x) = e^x, \qquad \ldots, \qquad f^{(n)}(x) = e^x, \qquad \ldots,$$

we have

$$f(0) = e^0 = 1, \qquad f'(0) = 1, \qquad \ldots, \qquad f^{(n)}(0) = 1, \qquad \ldots.$$

The Taylor series generated by f at $x = 0$ is

$$f(0) + f'(0)x + \frac{f''(0)}{2!}x^2 + \cdots + \frac{f^{(n)}(0)}{n!}x^n + \cdots$$

$$= 1 + x + \frac{x^2}{2} + \cdots + \frac{x^n}{n!} + \cdots$$

$$= \sum_{k=0}^{\infty} \frac{x^k}{k!}.$$

This is also the Maclaurin series for e^x. In Section 11.9 we will see that the series converges to e^x at every x.

The Taylor polynomial of order n at $x = 0$ is

$$P_n(x) = 1 + x + \frac{x^2}{2} + \cdots + \frac{x^n}{n!}.$$

See Figure 11.12. ■

EXAMPLE 3 Finding Taylor Polynomials for $\cos x$

Find the Taylor series and Taylor polynomials generated by $f(x) = \cos x$ at $x = 0$.

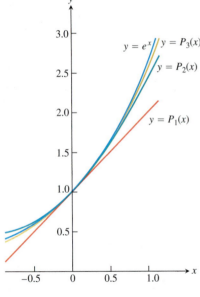

FIGURE 11.12 The graph of $f(x) = e^x$ and its Taylor polynomials
$P_1(x) = 1 + x$
$P_2(x) = 1 + x + (x^2/2!)$
$P_3(x) = 1 + x + (x^2/2!) + (x^3/3!)$.
Notice the very close agreement near the center $x = 0$ (Example 2).

Solution   The cosine and its derivatives are

$$f(x) = \cos x, \qquad f'(x) = -\sin x,$$
$$f''(x) = -\cos x, \qquad f^{(3)}(x) = \sin x,$$
$$\vdots \qquad\qquad\qquad \vdots$$
$$f^{(2n)}(x) = (-1)^n \cos x, \qquad f^{(2n+1)}(x) = (-1)^{n+1} \sin x.$$

At $x = 0$, the cosines are 1 and the sines are 0, so

$$f^{(2n)}(0) = (-1)^n, \qquad f^{(2n+1)}(0) = 0.$$

The Taylor series generated by f at 0 is

$$f(0) + f'(0)x + \frac{f''(0)}{2!}x^2 + \frac{f'''(0)}{3!}x^3 + \cdots + \frac{f^{(n)}(0)}{n!}x^n + \cdots$$
$$= 1 + 0 \cdot x - \frac{x^2}{2!} + 0 \cdot x^3 + \frac{x^4}{4!} + \cdots + (-1)^n \frac{x^{2n}}{(2n)!} + \cdots$$
$$= \sum_{k=0}^{\infty} \frac{(-1)^k x^{2k}}{(2k)!}.$$

This is also the Maclaurin series for $\cos x$. In Section 11.9, we will see that the series converges to $\cos x$ at every x.

Because $f^{(2n+1)}(0) = 0$, the Taylor polynomials of orders $2n$ and $2n + 1$ are identical:

$$P_{2n}(x) = P_{2n+1}(x) = 1 - \frac{x^2}{2!} + \frac{x^4}{4!} - \cdots + (-1)^n \frac{x^{2n}}{(2n)!}.$$

Figure 11.13 shows how well these polynomials approximate $f(x) = \cos x$ near $x = 0$. Only the right-hand portions of the graphs are given because the graphs are symmetric about the y-axis.   ■

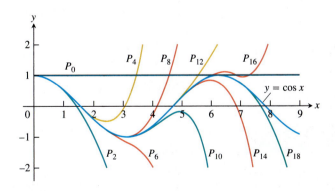

FIGURE 11.13   The polynomials

$$P_{2n}(x) = \sum_{k=0}^{n} \frac{(-1)^k x^{2k}}{(2k)!}$$

converge to $\cos x$ as $n \rightarrow \infty$. We can deduce the behavior of $\cos x$ arbitrarily far away solely from knowing the values of the cosine and its derivatives at $x = 0$ (Example 3).

EXAMPLE 4   A Function f Whose Taylor Series Converges at Every x but Converges to $f(x)$ Only at $x = 0$

It can be shown (though not easily) that

$$f(x) = \begin{cases} 0, & x = 0 \\ e^{-1/x^2}, & x \neq 0 \end{cases}$$

(Figure 11.14) has derivatives of all orders at $x = 0$ and that $f^{(n)}(0) = 0$ for all n. This means that the Taylor series generated by f at $x = 0$ is

$$f(0) + f'(0)x + \frac{f''(0)}{2!}x^2 + \cdots + \frac{f^{(n)}(0)}{n!}x^n + \cdots$$

$$= 0 + 0 \cdot x + 0 \cdot x^2 + \cdots + 0 \cdot x^n + \cdots$$

$$= 0 + 0 + \cdots + 0 + \cdots.$$

The series converges for every x (its sum is 0) but converges to $f(x)$ only at $x = 0$. ■

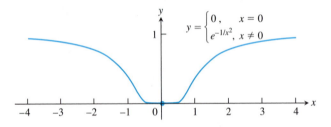

FIGURE 11.14 The graph of the continuous extension of $y = e^{-1/x^2}$ is so flat at the origin that all of its derivatives there are zero (Example 4).

Two questions still remain.

1. For what values of x can we normally expect a Taylor series to converge to its generating function?

2. How accurately do a function's Taylor polynomials approximate the function on a given interval?

The answers are provided by a theorem of Taylor in the next section.

EXERCISES 11.8

Finding Taylor Polynomials

In Exercises 1–8, find the Taylor polynomials of orders 0, 1, 2, and 3 generated by f at a.

1. $f(x) = \ln x, \quad a = 1$

2. $f(x) = \ln(1 + x), \quad a = 0$

3. $f(x) = 1/x, \quad a = 2$

4. $f(x) = 1/(x + 2), \quad a = 0$

5. $f(x) = \sin x, \quad a = \pi/4$

6. $f(x) = \cos x, \quad a = \pi/4$

7. $f(x) = \sqrt{x}, \quad a = 4$

8. $f(x) = \sqrt{x + 4}, \quad a = 0$

Finding Taylor Series at $x = 0$ (Maclaurin Series)

Find the Maclaurin series for the functions in Exercises 9–20.

9. e^{-x}

10. $e^{x/2}$

11. $\dfrac{1}{1 + x}$

12. $\dfrac{1}{1 - x}$

13. $\sin 3x$

14. $\sin \dfrac{x}{2}$

15. $7 \cos(-x)$

16. $5 \cos \pi x$

17. $\cosh x = \dfrac{e^x + e^{-x}}{2}$

18. $\sinh x = \dfrac{e^x - e^{-x}}{2}$

19. $x^4 - 2x^3 - 5x + 4$

20. $(x + 1)^2$

Finding Taylor Series

In Exercises 21–28, find the Taylor series generated by f at $x = a$.

21. $f(x) = x^3 - 2x + 4, \quad a = 2$

22. $f(x) = 2x^3 + x^2 + 3x - 8, \quad a = 1$

23. $f(x) = x^4 + x^2 + 1, \quad a = -2$

24. $f(x) = 3x^5 - x^4 + 2x^3 + x^2 - 2, \quad a = -1$

25. $f(x) = 1/x^2, \quad a = 1$

26. $f(x) = x/(1 - x), \quad a = 0$

27. $f(x) = e^x, \quad a = 2$

28. $f(x) = 2^x, \quad a = 1$

Theory and Examples

29. Use the Taylor series generated by e^x at $x = a$ to show that

$$e^x = e^a \left[1 + (x - a) + \frac{(x - a)^2}{2!} + \cdots \right].$$

30. (*Continuation of Exercise 29.*) Find the Taylor series generated by e^x at $x = 1$. Compare your answer with the formula in Exercise 29.

31. Let $f(x)$ have derivatives through order n at $x = a$. Show that the Taylor polynomial of order n and its first n derivatives have the same values that f and its first n derivatives have at $x = a$.

32. Of all polynomials of degree $\leq n$, the Taylor polynomial of order n gives the best approximation   Suppose that $f(x)$ is differentiable on an interval centered at $x = a$ and that $g(x) = b_0 + b_1(x - a) + \cdots + b_n(x - a)^n$ is a polynomial of degree n with constant coefficients $b_0, \ldots, b_n$. Let $E(x) = f(x) - g(x)$. Show that if we impose on g the conditions

a. $E(a) = 0$ The approximation error is zero at $x = a$.

b. $\displaystyle \lim_{x \to a} \frac{E(x)}{(x - a)^n} = 0,$ The error is negligible when compared to $(x - a)^n$.

then

$$g(x) = f(a) + f'(a)(x - a) + \frac{f''(a)}{2!}(x - a)^2 + \cdots$$
$$+ \frac{f^{(n)}(a)}{n!}(x - a)^n.$$

Thus, the Taylor polynomial $P_n(x)$ is the only polynomial of degree less than or equal to n whose error is both zero at $x = a$ and negligible when compared with $(x - a)^n$.

Quadratic Approximations

The Taylor polynomial of order 2 generated by a twice-differentiable function $f(x)$ at $x = a$ is called the **quadratic approximation** of f at $x = a$. In Exercises 33–38, find the **(a)** linearization (Taylor polynomial of order 1) and **(b)** quadratic approximation of f at $x = 0$.

33. $f(x) = \ln(\cos x)$

34. $f(x) = e^{\sin x}$

35. $f(x) = 1/\sqrt{1 - x^2}$

36. $f(x) = \cosh x$

37. $f(x) = \sin x$

38. $f(x) = \tan x$

11.9   Convergence of Taylor Series; Error Estimates

This section addresses the two questions left unanswered by Section 11.8:

1.   When does a Taylor series converge to its generating function?

2.   How accurately do a function's Taylor polynomials approximate the function on a given interval?

Taylor's Theorem

We answer these questions with the following theorem.

THEOREM 22   Taylor's Theorem

If f and its first n derivatives f', f'', ..., $f^{(n)}$ are continuous on the closed interval between a and b, and $f^{(n)}$ is differentiable on the open interval between a and b, then there exists a number c between a and b such that

$$f(b) = f(a) + f'(a)(b - a) + \frac{f''(a)}{2!}(b - a)^2 + \cdots$$

$$+ \frac{f^{(n)}(a)}{n!}(b - a)^n + \frac{f^{(n+1)}(c)}{(n + 1)!}(b - a)^{n+1}.$$

Taylor's Theorem is a generalization of the Mean Value Theorem (Exercise 39). There is a proof of Taylor's Theorem at the end of this section.

When we apply Taylor's Theorem, we usually want to hold a fixed and treat b as an independent variable. Taylor's formula is easier to use in circumstances like these if we change b to x. Here is a version of the theorem with this change.

Taylor's Formula

If f has derivatives of all orders in an open interval I containing a, then for each positive integer n and for each x in I,

$$f(x) = f(a) + f'(a)(x - a) + \frac{f''(a)}{2!}(x - a)^2 + \cdots$$

$$+ \frac{f^{(n)}(a)}{n!}(x - a)^n + R_n(x), \tag{1}$$

where

$$R_n(x) = \frac{f^{(n+1)}(c)}{(n + 1)!}(x - a)^{n+1} \qquad \text{for some } c \text{ between } a \text{ and } x. \tag{2}$$

When we state Taylor's theorem this way, it says that for each $x \in I$,

$$f(x) = P_n(x) + R_n(x).$$

The function $R_n(x)$ is determined by the value of the $(n + 1)$st derivative $f^{(n+1)}$ at a point c that depends on both a and x, and which lies somewhere between them. For any value of n we want, the equation gives both a polynomial approximation of f of that order and a formula for the error involved in using that approximation over the interval I.

Equation (1) is called **Taylor's formula**. The function $R_n(x)$ is called the **remainder of order n** or the **error term** for the approximation of f by $P_n(x)$ over I. If $R_n(x) \to 0$ as $n \to \infty$ for all $x \in I$, we say that the Taylor series generated by f at $x = a$ **converges** to f on I, and we write

$$f(x) = \sum_{k=0}^{\infty} \frac{f^{(k)}(a)}{k!}(x - a)^k.$$

Often we can estimate R_n without knowing the value of c, as the following example illustrates.

EXAMPLE 1   The Taylor Series for e^x Revisited

Show that the Taylor series generated by $f(x) = e^x$ at $x = 0$ converges to $f(x)$ for every real value of x.

Solution   The function has derivatives of all orders throughout the interval $I = (-\infty, \infty)$. Equations (1) and (2) with $f(x) = e^x$ and $a = 0$ give

$$e^x = 1 + x + \frac{x^2}{2!} + \cdots + \frac{x^n}{n!} + R_n(x) \qquad \text{Polynomial from Section 11.8, Example 2}$$

and

$$R_n(x) = \frac{e^c}{(n+1)!} x^{n+1} \qquad \text{for some } c \text{ between } 0 \text{ and } x.$$

Since e^x is an increasing function of x, e^c lies between $e^0 = 1$ and e^x. When x is negative, so is c, and $e^c < 1$. When x is zero, $e^x = 1$ and $R_n(x) = 0$. When x is positive, so is c, and $e^c < e^x$. Thus,

$$|R_n(x)| \leq \frac{|x|^{n+1}}{(n+1)!} \qquad \text{when } x \leq 0,$$

and

$$|R_n(x)| < e^x \frac{x^{n+1}}{(n+1)!} \qquad \text{when } x > 0.$$

Finally, because

$$\lim_{n \to \infty} \frac{x^{n+1}}{(n+1)!} = 0 \qquad \text{for every } x, \qquad \text{Section 11.1}$$

$\lim_{n \to \infty} R_n(x) = 0$, and the series converges to e^x for every x. Thus,

$$e^x = \sum_{k=0}^{\infty} \frac{x^k}{k!} = 1 + x + \frac{x^2}{2!} + \cdots + \frac{x^k}{k!} + \cdots. \tag{3}$$

∎

Estimating the Remainder

It is often possible to estimate $R_n(x)$ as we did in Example 1. This method of estimation is so convenient that we state it as a theorem for future reference.

THEOREM 23   The Remainder Estimation Theorem

If there is a positive constant M such that $|f^{(n+1)}(t)| \leq M$ for all t between x and a, inclusive, then the remainder term $R_n(x)$ in Taylor's Theorem satisfies the inequality

$$|R_n(x)| \leq M \frac{|x - a|^{n+1}}{(n+1)!}.$$

If this condition holds for every n and the other conditions of Taylor's Theorem are satisfied by f, then the series converges to $f(x)$.

We are now ready to look at some examples of how the Remainder Estimation Theorem and Taylor's Theorem can be used together to settle questions of convergence. As you will see, they can also be used to determine the accuracy with which a function is approximated by one of its Taylor polynomials.

EXAMPLE 2   The Taylor Series for sin x at $x = 0$

Show that the Taylor series for sin x at $x = 0$ converges for all x.

Solution   The function and its derivatives are

$$f(x) = \quad \sin x, \qquad\qquad f'(x) = \quad \cos x,$$

$$f''(x) = \quad -\sin x, \qquad\qquad f'''(x) = \quad -\cos x,$$

$$\vdots \qquad\qquad\qquad\qquad \vdots$$

$$f^{(2k)}(x) = (-1)^k \sin x, \qquad f^{(2k+1)}(x) = (-1)^k \cos x,$$

so

$$f^{(2k)}(0) = 0 \quad\text{and}\quad f^{(2k+1)}(0) = (-1)^k.$$

The series has only odd-powered terms and, for $n = 2k + 1$, Taylor's Theorem gives

$$\sin x = x - \frac{x^3}{3!} + \frac{x^5}{5!} - \cdots + \frac{(-1)^k x^{2k+1}}{(2k + 1)!} + R_{2k+1}(x).$$

All the derivatives of sin x have absolute values less than or equal to 1, so we can apply the Remainder Estimation Theorem with $M = 1$ to obtain

$$|R_{2k+1}(x)| \le 1 \cdot \frac{|x|^{2k+2}}{(2k + 2)!}.$$

Since $(|x|^{2k+2}/(2k + 2)!) \to 0$ as $k \to \infty$, whatever the value of x, $R_{2k+1}(x) \to 0$, and the Maclaurin series for sin x converges to sin x for every x. Thus,

$$\sin x = \sum_{k=0}^{\infty} \frac{(-1)^k x^{2k+1}}{(2k + 1)!} = x - \frac{x^3}{3!} + \frac{x^5}{5!} - \frac{x^7}{7!} + \cdots. \tag{4}$$

■

EXAMPLE 3   The Taylor Series for cos x at $x = 0$ Revisited

Show that the Taylor series for cos x at $x = 0$ converges to cos x for every value of x.

Solution   We add the remainder term to the Taylor polynomial for cos x (Section 11.8, Example 3) to obtain Taylor's formula for cos x with $n = 2k$:

$$\cos x = 1 - \frac{x^2}{2!} + \frac{x^4}{4!} - \cdots + (-1)^k \frac{x^{2k}}{(2k)!} + R_{2k}(x).$$

Because the derivatives of the cosine have absolute value less than or equal to 1, the Remainder Estimation Theorem with $M = 1$ gives

$$|R_{2k}(x)| \leq 1 \cdot \frac{|x|^{2k+1}}{(2k + 1)!}.$$

For every value of x, $R_{2k} \to 0$ as $k \to \infty$. Therefore, the series converges to $\cos x$ for every value of x. Thus,

$$\cos x = \sum_{k=0}^{\infty} \frac{(-1)^k x^{2k}}{(2k)!} = 1 - \frac{x^2}{2!} + \frac{x^4}{4!} - \frac{x^6}{6!} + \cdots. \tag{5}$$

$\blacksquare$

EXAMPLE 4   Finding a Taylor Series by Substitution

Find the Taylor series for $\cos 2x$ at $x = 0$.

Solution   We can find the Taylor series for $\cos 2x$ by substituting $2x$ for x in the Taylor series for $\cos x$:

$$\cos 2x = \sum_{k=0}^{\infty} \frac{(-1)^k (2x)^{2k}}{(2k)!} = 1 - \frac{(2x)^2}{2!} + \frac{(2x)^4}{4!} - \frac{(2x)^6}{6!} + \cdots \qquad \text{Equation (5)}$$
$$\text{with } 2x \text{ for } x$$

$$= 1 - \frac{2^2 x^2}{2!} + \frac{2^4 x^4}{4!} - \frac{2^6 x^6}{6!} + \cdots$$

$$= \sum_{k=0}^{\infty} (-1)^k \frac{2^{2k} x^{2k}}{(2k)!}.$$

Equation (5) holds for $-\infty < x < \infty$, implying that it holds for $-\infty < 2x < \infty$, so the newly created series converges for all x. Exercise 45 explains why the series is in fact the Taylor series for $\cos 2x$. $\blacksquare$

EXAMPLE 5   Finding a Taylor Series by Multiplication

Find the Taylor series for $x \sin x$ at $x = 0$.

Solution   We can find the Taylor series for $x \sin x$ by multiplying the Taylor series for $\sin x$ (Equation 4) by x:

$$x \sin x = x \left(x - \frac{x^3}{3!} + \frac{x^5}{5!} - \frac{x^7}{7!} + \cdots \right)$$

$$= x^2 - \frac{x^4}{3!} + \frac{x^6}{5!} - \frac{x^8}{7!} + \cdots.$$

The new series converges for all x because the series for $\sin x$ converges for all x. Exercise 45 explains why the series is the Taylor series for $x \sin x$. $\blacksquare$

Truncation Error

The Taylor series for e^x at $x = 0$ converges to e^x for all x. But we still need to decide how many terms to use to approximate e^x to a given degree of accuracy. We get this information from the Remainder Estimation Theorem.

EXAMPLE 6   Calculate e with an error of less than 10^{-6}.

Solution   We can use the result of Example 1 with $x = 1$ to write

$$e = 1 + 1 + \frac{1}{2!} + \cdots + \frac{1}{n!} + R_n(1),$$

with

$$R_n(1) = e^c \frac{1}{(n + 1)!} \qquad \text{for some } c \text{ between 0 and 1.}$$

For the purposes of this example, we assume that we know that $e < 3$. Hence, we are certain that

$$\frac{1}{(n + 1)!} < R_n(1) < \frac{3}{(n + 1)!}$$

because $1 < e^c < 3$ for $0 < c < 1$.

By experiment we find that $1/9! > 10^{-6}$, while $3/10! < 10^{-6}$. Thus we should take $(n + 1)$ to be at least 10, or n to be at least 9. With an error of less than 10^{-6},

$$e = 1 + 1 + \frac{1}{2} + \frac{1}{3!} + \cdots + \frac{1}{9!} \approx 2.718282. \qquad \blacksquare$$

EXAMPLE 7   For what values of x can we replace $\sin x$ by $x - (x^3/3!)$ with an error of magnitude no greater than 3×10^{-4}?

Solution   Here we can take advantage of the fact that the Taylor series for $\sin x$ is an alternating series for every nonzero value of x. According to the Alternating Series Estimation Theorem (Section 11.6), the error in truncating

$$\sin x = x - \frac{x^3}{3!} \Big| + \frac{x^5}{5!} - \cdots$$

after $(x^3/3!)$ is no greater than

$$\left| \frac{x^5}{5!} \right| = \frac{|x|^5}{120}.$$

Therefore the error will be less than or equal to 3×10^{-4} if

$$\frac{|x|^5}{120} < 3 \times 10^{-4} \qquad \text{or} \qquad |x| < \sqrt[5]{360 \times 10^{-4}} \approx 0.514. \qquad \begin{array}{l} \text{\color{blue}{Rounded down,}} \\ \text{\color{blue}{to be safe}} \end{array}$$

The Alternating Series Estimation Theorem tells us something that the Remainder Estimation Theorem does not: namely, that the estimate $x - (x^3/3!)$ for $\sin x$ is an underestimate when x is positive because then $x^5/120$ is positive.

Figure 11.15 shows the graph of $\sin x$, along with the graphs of a number of its approximating Taylor polynomials. The graph of $P_3(x) = x - (x^3/3!)$ is almost indistinguishable from the sine curve when $-1 \le x \le 1$.

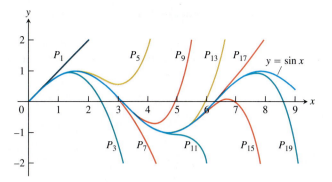

FIGURE 11.15 The polynomials

$$P_{2n+1}(x) = \sum_{k=0}^{n} \frac{(-1)^k x^{2k+1}}{(2k+1)!}$$

converge to $\sin x$ as $n \to \infty$. Notice how closely $P_3(x)$ approximates the sine curve for $x < 1$ (Example 7).

You might wonder how the estimate given by the Remainder Estimation Theorem compares with the one just obtained from the Alternating Series Estimation Theorem. If we write

$$\sin x = x - \frac{x^3}{3!} + R_3,$$

then the Remainder Estimation Theorem gives

$$|R_3| \le 1 \cdot \frac{|x|^4}{4!} = \frac{|x|^4}{24},$$

which is not as good. But if we recognize that $x - (x^3/3!) = 0 + x + 0x^2 - (x^3/3!) + 0x^4$ is the Taylor polynomial of order 4 as well as of order 3, then

$$\sin x = x - \frac{x^3}{3!} + 0 + R_4,$$

and the Remainder Estimation Theorem with $M = 1$ gives

$$|R_4| \le 1 \cdot \frac{|x|^5}{5!} = \frac{|x|^5}{120}.$$

This is what we had from the Alternating Series Estimation Theorem. ∎

Combining Taylor Series

On the intersection of their intervals of convergence, Taylor series can be added, subtracted, and multiplied by constants, and the results are once again Taylor series. The Taylor series for $f(x) + g(x)$ is the sum of the Taylor series for $f(x)$ and $g(x)$ because the nth derivative of $f + g$ is $f^{(n)} + g^{(n)}$, and so on. Thus we obtain the Taylor series for $(1 + \cos 2x)/2$ by adding 1 to the Taylor series for $\cos 2x$ and dividing the combined results by 2, and the Taylor series for $\sin x + \cos x$ is the term-by-term sum of the Taylor series for $\sin x$ and $\cos x$.

Euler's Identity

As you may recall, a complex number is a number of the form $a + bi$, where a and b are real numbers and $i = \sqrt{-1}$. If we substitute $x = i\theta$ (θ real) in the Taylor series for e^x and use the relations

$$i^2 = -1, \qquad i^3 = i^2i = -i, \qquad i^4 = i^2i^2 = 1, \qquad i^5 = i^4i = i,$$

and so on, to simplify the result, we obtain

$$e^{i\theta} = 1 + \frac{i\theta}{1!} + \frac{i^2\theta^2}{2!} + \frac{i^3\theta^3}{3!} + \frac{i^4\theta^4}{4!} + \frac{i^5\theta^5}{5!} + \frac{i^6\theta^6}{6!} + \cdots$$

$$= \left(1 - \frac{\theta^2}{2!} + \frac{\theta^4}{4!} - \frac{\theta^6}{6!} + \cdots\right) + i\left(\theta - \frac{\theta^3}{3!} + \frac{\theta^5}{5!} - \cdots\right) = \cos\theta + i\sin\theta.$$

This does not *prove* that $e^{i\theta} = \cos\theta + i\sin\theta$ because we have not yet defined what it means to raise e to an imaginary power. Rather, it says how to define $e^{i\theta}$ to be consistent with other things we know.

> **DEFINITION**
>
> For any real number θ, $e^{i\theta} = \cos\theta + i\sin\theta$.   (6)

Equation (6), called **Euler's identity**, enables us to define e^{a+bi} to be $e^a \cdot e^{bi}$ for any complex number $a + bi$. One consequence of the identity is the equation

$$e^{i\pi} = -1.$$

When written in the form $e^{i\pi} + 1 = 0$, this equation combines five of the most important constants in mathematics.

A Proof of Taylor's Theorem

We prove Taylor's theorem assuming $a < b$. The proof for $a > b$ is nearly the same.

The Taylor polynomial

$$P_n(x) = f(a) + f'(a)(x - a) + \frac{f''(a)}{2!}(x - a)^2 + \cdots + \frac{f^{(n)}(a)}{n!}(x - a)^n$$

and its first n derivatives match the function f and its first n derivatives at $x = a$. We do not disturb that matching if we add another term of the form $K(x - a)^{n+1}$, where K is any constant, because such a term and its first n derivatives are all equal to zero at $x = a$. The new function

$$\phi_n(x) = P_n(x) + K(x - a)^{n+1}$$

and its first n derivatives still agree with f and its first n derivatives at $x = a$.

We now choose the particular value of K that makes the curve $y = \phi_n(x)$ agree with the original curve $y = f(x)$ at $x = b$. In symbols,

$$f(b) = P_n(b) + K(b - a)^{n+1}, \qquad \text{or} \qquad K = \frac{f(b) - P_n(b)}{(b - a)^{n+1}}. \qquad (7)$$

With K defined by Equation (7), the function

$$F(x) = f(x) - \phi_n(x)$$

measures the difference between the original function f and the approximating function ϕ_n for each x in $[a, b]$.

We now use Rolle's Theorem (Section 4.2). First, because $F(a) = F(b) = 0$ and both F and F' are continuous on $[a, b]$, we know that

$$F'(c_1) = 0 \quad \text{for some } c_1 \text{ in } (a, b).$$

Next, because $F'(a) = F'(c_1) = 0$ and both F' and F'' are continuous on $[a, c_1]$, we know that

$$F''(c_2) = 0 \quad \text{for some } c_2 \text{ in } (a, c_1).$$

Rolle's Theorem, applied successively to $F'', F''', \ldots, F^{(n-1)}$ implies the existence of

$$c_3 \quad \text{in } (a, c_2) \quad \text{such that } F'''(c_3) = 0,$$
$$c_4 \quad \text{in } (a, c_3) \quad \text{such that } F^{(4)}(c_4) = 0,$$
$$\vdots$$
$$c_n \quad \text{in } (a, c_{n-1}) \quad \text{such that } F^{(n)}(c_n) = 0.$$

Finally, because $F^{(n)}$ is continuous on $[a, c_n]$ and differentiable on (a, c_n), and $F^{(n)}(a) = F^{(n)}(c_n) = 0$, Rolle's Theorem implies that there is a number c_{n+1} in (a, c_n) such that

$$F^{(n+1)}(c_{n+1}) = 0. \tag{8}$$

If we differentiate $F(x) = f(x) - P_n(x) - K(x - a)^{n+1}$ a total of $n + 1$ times, we get

$$F^{(n+1)}(x) = f^{(n+1)}(x) - 0 - (n + 1)!K. \tag{9}$$

Equations (8) and (9) together give

$$K = \frac{f^{(n+1)}(c)}{(n + 1)!} \quad \text{for some number } c = c_{n+1} \text{ in } (a, b). \tag{10}$$

Equations (7) and (10) give

$$f(b) = P_n(b) + \frac{f^{(n+1)}(c)}{(n + 1)!}(b - a)^{n+1}.$$

This concludes the proof. ■

EXERCISES 11.9

Taylor Series by Substitution

Use substitution (as in Example 4) to find the Taylor series at $x = 0$ of the functions in Exercises 1–6.

1. e^{-5x}
2. $e^{-x/2}$
3. $5\sin(-x)$
4. $\sin\left(\dfrac{\pi x}{2}\right)$
5. $\cos\sqrt{x + 1}$
6. $\cos\left(x^{3/2}/\sqrt{2}\right)$

More Taylor Series

Find Taylor series at $x = 0$ for the functions in Exercises 7–18.

7. xe^x
8. $x^2\sin x$
9. $\dfrac{x^2}{2} - 1 + \cos x$
10. $\sin x - x + \dfrac{x^3}{3!}$
11. $x\cos \pi x$
12. $x^2\cos(x^2)$

13. $\cos^2 x$ (*Hint:* $\cos^2 x = (1 + \cos 2x)/2$.)

14. $\sin^2 x$ **15.** $\dfrac{x^2}{1 - 2x}$ **16.** $x \ln(1 + 2x)$

17. $\dfrac{1}{(1 - x)^2}$ **18.** $\dfrac{2}{(1 - x)^3}$

Error Estimates

19. For approximately what values of x can you replace $\sin x$ by $x - (x^3/6)$ with an error of magnitude no greater than 5×10^{-4}? Give reasons for your answer.

20. If $\cos x$ is replaced by $1 - (x^2/2)$ and $|x| < 0.5$, what estimate can be made of the error? Does $1 - (x^2/2)$ tend to be too large, or too small? Give reasons for your answer.

21. How close is the approximation $\sin x = x$ when $|x| < 10^{-3}$? For which of these values of x is $x < \sin x$?

22. The estimate $\sqrt{1 + x} = 1 + (x/2)$ is used when x is small. Estimate the error when $|x| < 0.01$.

23. The approximation $e^x = 1 + x + (x^2/2)$ is used when x is small. Use the Remainder Estimation Theorem to estimate the error when $|x| < 0.1$.

24. (*Continuation of Exercise 23.*) When $x < 0$, the series for e^x is an alternating series. Use the Alternating Series Estimation Theorem to estimate the error that results from replacing e^x by $1 + x + (x^2/2)$ when $-0.1 < x < 0$. Compare your estimate with the one you obtained in Exercise 23.

25. Estimate the error in the approximation $\sinh x = x + (x^3/3!)$ when $|x| < 0.5$. (*Hint:* Use R_4, not R_3.)

26. When $0 \le h \le 0.01$, show that e^h may be replaced by $1 + h$ with an error of magnitude no greater than 0.6% of h. Use $e^{0.01} = 1.01$.

27. For what positive values of x can you replace $\ln(1 + x)$ by x with an error of magnitude no greater than 1% of the value of x?

28. You plan to estimate $\pi/4$ by evaluating the Maclaurin series for $\tan^{-1} x$ at $x = 1$. Use the Alternating Series Estimation Theorem to determine how many terms of the series you would have to add to be sure the estimate is good to two decimal places.

29. a. Use the Taylor series for $\sin x$ and the Alternating Series Estimation Theorem to show that

$$1 - \frac{x^2}{6} < \frac{\sin x}{x} < 1, \quad x \ne 0.$$

T **b.** Graph $f(x) = (\sin x)/x$ together with the functions $y = 1 - (x^2/6)$ and $y = 1$ for $-5 \le x \le 5$. Comment on the relationships among the graphs.

30. a. Use the Taylor series for $\cos x$ and the Alternating Series Estimation Theorem to show that

$$\frac{1}{2} - \frac{x^2}{24} < \frac{1 - \cos x}{x^2} < \frac{1}{2}, \quad x \ne 0.$$

(This is the inequality in Section 2.2, Exercise 52.)

T **b.** Graph $f(x) = (1 - \cos x)/x^2$ together with $y = (1/2) - (x^2/24)$ and $y = 1/2$ for $-9 \le x \le 9$. Comment on the relationships among the graphs.

Finding and Identifying Maclaurin Series

Recall that the Maclaurin series is just another name for the Taylor series at $x = 0$. Each of the series in Exercises 31–34 is the value of the Maclaurin series of a function $f(x)$ at some point. What function and what point? What is the sum of the series?

31. $(0.1) - \dfrac{(0.1)^3}{3!} + \dfrac{(0.1)^5}{5!} - \cdots + \dfrac{(-1)^k(0.1)^{2k+1}}{(2k + 1)!} + \cdots$

32. $1 - \dfrac{\pi^2}{4^2 \cdot 2!} + \dfrac{\pi^4}{4^4 \cdot 4!} - \cdots + \dfrac{(-1)^k(\pi)^{2k}}{4^{2k} \cdot (2k!)} + \cdots$

33. $\dfrac{\pi}{3} - \dfrac{\pi^3}{3^3 \cdot 3} + \dfrac{\pi^5}{3^5 \cdot 5} - \cdots + \dfrac{(-1)^k \pi^{2k+1}}{3^{2k+1}(2k + 1)} + \cdots$

34. $\pi - \dfrac{\pi^2}{2} + \dfrac{\pi^3}{3} - \cdots + (-1)^{k-1}\dfrac{\pi^k}{k} + \cdots$

35. Multiply the Maclaurin series for e^x and $\sin x$ together to find the first five nonzero terms of the Maclaurin series for $e^x \sin x$.

36. Multiply the Maclaurin series for e^x and $\cos x$ together to find the first five nonzero terms of the Maclaurin series for $e^x \cos x$.

37. Use the identity $\sin^2 x = (1 - \cos 2x)/2$ to obtain the Maclaurin series for $\sin^2 x$. Then differentiate this series to obtain the Maclaurin series for $2 \sin x \cos x$. Check that this is the series for $\sin 2x$.

38. (*Continuation of Exercise 37.*) Use the identity $\cos^2 x = \cos 2x + \sin^2 x$ to obtain a power series for $\cos^2 x$.

Theory and Examples

39. Taylor's Theorem and the Mean Value Theorem Explain how the Mean Value Theorem (Section 4.2, Theorem 4) is a special case of Taylor's Theorem.

40. Linearizations at inflection points Show that if the graph of a twice-differentiable function $f(x)$ has an inflection point at $x = a$, then the linearization of f at $x = a$ is also the quadratic approximation of f at $x = a$. This explains why tangent lines fit so well at inflection points.

41. The (second) second derivative test Use the equation

$$f(x) = f(a) + f'(a)(x - a) + \frac{f''(c_2)}{2}(x - a)^2$$

to establish the following test.

    Let f have continuous first and second derivatives and suppose that $f'(a) = 0$. Then

a. f has a local maximum at a if $f'' \le 0$ throughout an interval whose interior contains a;

b. f has a local minimum at a if $f'' \ge 0$ throughout an interval whose interior contains a.

42. A cubic approximation Use Taylor's formula with $a = 0$ and $n = 3$ to find the standard cubic approximation of $f(x) = 1/(1 - x)$ at $x = 0$. Give an upper bound for the magnitude of the error in the approximation when $|x| \le 0.1$.

43. a. Use Taylor's formula with $n = 2$ to find the quadratic approximation of $f(x) = (1 + x)^k$ at $x = 0$ (k a constant).

   b. If $k = 3$, for approximately what values of x in the interval $[0, 1]$ will the error in the quadratic approximation be less than $1/100$?

44. Improving approximations to π

   a. Let P be an approximation of π accurate to n decimals. Show that $P + \sin P$ gives an approximation correct to $3n$ decimals. (*Hint:* Let $P = \pi + x$.)

   T b. Try it with a calculator.

45. The Taylor series generated by $f(x) = \sum_{n=0}^{\infty} a_n x^n$ **is** $\sum_{n=0}^{\infty} a_n x^n$ A function defined by a power series $\sum_{n=0}^{\infty} a_n x^n$ with a radius of convergence $c > 0$ has a Taylor series that converges to the function at every point of $(-c, c)$. Show this by showing that the Taylor series generated by $f(x) = \sum_{n=0}^{\infty} a_n x^n$ is the series $\sum_{n=0}^{\infty} a_n x^n$ itself.

   An immediate consequence of this is that series like

$$x \sin x = x^2 - \frac{x^4}{3!} + \frac{x^6}{5!} - \frac{x^8}{7!} + \cdots$$

and

$$x^2 e^x = x^2 + x^3 + \frac{x^4}{2!} + \frac{x^5}{3!} + \cdots,$$

obtained by multiplying Taylor series by powers of x, as well as series obtained by integration and differentiation of convergent power series, are themselves the Taylor series generated by the functions they represent.

46. Taylor series for even functions and odd functions (*Continuation of Section 11.7, Exercise 45.*) Suppose that $f(x) = \sum_{n=0}^{\infty} a_n x^n$ converges for all x in an open interval $(-c, c)$. Show that

   a. If f is even, then $a_1 = a_3 = a_5 = \cdots = 0$, i.e., the Taylor series for f at $x = 0$ contains only even powers of x.

   b. If f is odd, then $a_0 = a_2 = a_4 = \cdots = 0$, i.e., the Taylor series for f at $x = 0$ contains only odd powers of x.

47. Taylor polynomials of periodic functions

   a. Show that every continuous periodic function $f(x)$, $-\infty < x < \infty$, is bounded in magnitude by showing that there exists a positive constant M such that $|f(x)| \le M$ for all x.

   b. Show that the graph of every Taylor polynomial of positive degree generated by $f(x) = \cos x$ must eventually move away from the graph of $\cos x$ as $|x|$ increases. You can see this in Figure 11.13. The Taylor polynomials of $\sin x$ behave in a similar way (Figure 11.15).

T 48. a. Graph the curves $y = (1/3) - (x^2)/5$ and $y = (x - \tan^{-1} x)/x^3$ together with the line $y = 1/3$.

   b. Use a Taylor series to explain what you see. What is

$$\lim_{x \to 0} \frac{x - \tan^{-1} x}{x^3} ?$$

Euler's Identity

49. Use Equation (6) to write the following powers of e in the form $a + bi$.

   a. $e^{-i\pi}$ **b.** $e^{i\pi/4}$ **c.** $e^{-i\pi/2}$

50. Use Equation (6) to show that

$$\cos\theta = \frac{e^{i\theta} + e^{-i\theta}}{2} \quad \text{and} \quad \sin\theta = \frac{e^{i\theta} - e^{-i\theta}}{2i}.$$

51. Establish the equations in Exercise 50 by combining the formal Taylor series for $e^{i\theta}$ and $e^{-i\theta}$.

52. Show that

   a. $\cosh i\theta = \cos\theta$,   **b.** $\sinh i\theta = i \sin\theta$.

53. By multiplying the Taylor series for e^x and $\sin x$, find the terms through x^5 of the Taylor series for $e^x \sin x$. This series is the imaginary part of the series for

$$e^x \cdot e^{ix} = e^{(1+i)x}.$$

Use this fact to check your answer. For what values of x should the series for $e^x \sin x$ converge?

54. When a and b are real, we define $e^{(a+ib)x}$ with the equation

$$e^{(a+ib)x} = e^{ax} \cdot e^{ibx} = e^{ax}(\cos bx + i \sin bx).$$

Differentiate the right-hand side of this equation to show that

$$\frac{d}{dx} e^{(a+ib)x} = (a + ib)e^{(a+ib)x}.$$

Thus the familiar rule $(d/dx)e^{kx} = ke^{kx}$ holds for k complex as well as real.

55. Use the definition of $e^{i\theta}$ to show that for any real numbers $\theta, \theta_1,$ and θ_2,

   a. $e^{i\theta_1}e^{i\theta_2} = e^{i(\theta_1+\theta_2)}$,   **b.** $e^{-i\theta} = 1/e^{i\theta}$.

56. Two complex numbers $a + ib$ and $c + id$ are equal if and only if $a = c$ and $b = d$. Use this fact to evaluate

$$\int e^{ax} \cos bx \, dx \quad \text{and} \quad \int e^{ax} \sin bx \, dx$$

from

$$\int e^{(a+ib)x} \, dx = \frac{a - ib}{a^2 + b^2} e^{(a+ib)x} + C,$$

where $C = C_1 + iC_2$ is a complex constant of integration.

Linear, Quadratic, and Cubic Approximations

Taylor's formula with $n = 1$ and $a = 0$ gives the linearization of a function at $x = 0$. With $n = 2$ and $n = 3$ we obtain the standard quadratic and cubic approximations. In these exercises we explore the errors associated with these approximations. We seek answers to two questions:

a. For what values of x can the function be replaced by each approximation with an error less than 10^{-2}?

b. What is the maximum error we could expect if we replace the function by each approximation over the specified interval?

Using a CAS, perform the following steps to aid in answering questions (a) and (b) for the functions and intervals in Exercises 57–62.

Step 1: Plot the function over the specified interval.

Step 2: Find the Taylor polynomials $P_1(x)$, $P_2(x)$, and $P_3(x)$ at $x = 0$.

Step 3: Calculate the $(n + 1)$st derivative $f^{(n+1)}(c)$ associated with the remainder term for each Taylor polynomial. Plot the derivative as a function of c over the specified interval and estimate its maximum absolute value, M.

Step 4: Calculate the remainder $R_n(x)$ for each polynomial. Using the estimate M from Step 3 in place of $f^{(n+1)}(c)$, plot $R_n(x)$ over the specified interval. Then estimate the values of x that answer question (a).

Step 5: Compare your estimated error with the actual error $E_n(x) = |f(x) - P_n(x)|$ by plotting $E_n(x)$ over the specified interval. This will help answer question (b).

Step 6: Graph the function and its three Taylor approximations together. Discuss the graphs in relation to the information discovered in Steps 4 and 5.

57. $f(x) = \dfrac{1}{\sqrt{1 + x}}, \quad |x| \le \dfrac{3}{4}$

58. $f(x) = (1 + x)^{3/2}, \quad -\dfrac{1}{2} \le x \le 2$

59. $f(x) = \dfrac{x}{x^2 + 1}, \quad |x| \le 2$

60. $f(x) = (\cos x)(\sin 2x), \quad |x| \le 2$

61. $f(x) = e^{-x} \cos 2x, \quad |x| \le 1$

62. $f(x) = e^{x/3} \sin 2x, \quad |x| \le 2$

11.10 Applications of Power Series

This section introduces the binomial series for estimating powers and roots and shows how series are sometimes used to approximate the solution of an initial value problem, to evaluate nonelementary integrals, and to evaluate limits that lead to indeterminate forms. We provide a self-contained derivation of the Taylor series for $\tan^{-1} x$ and conclude with a reference table of frequently used series.

The Binomial Series for Powers and Roots

The Taylor series generated by $f(x) = (1 + x)^m$, when m is constant, is

$$1 + mx + \frac{m(m - 1)}{2!}x^2 + \frac{m(m - 1)(m - 2)}{3!}x^3 + \cdots$$

$$+ \frac{m(m - 1)(m - 2)\cdots(m - k + 1)}{k!}x^k + \cdots. \qquad (1)$$

This series, called the **binomial series**, converges absolutely for $|x| < 1$. To derive the

Best Practices for Supporting Children's Mental Health

How to tell if a child is emotionally healthy, pp. 122–125

Tips for cultivating an emotionally positive climate in the children's center, pp. 126–128

Practical ways to help children achieve healthy emotional development, pp. 129–133

What to do when children are emotionally upset, pp. 133–134

How to help children achieve competence, pp. 136–138

Tips for helping families deal with a crisis, pp. 145–147

Tips for helping a child deal with a crisis, pp. 147–151

Suggestions for helping children cope with a disaster, p. 157

Suggestions for helping children in frightening times—How to assist children in a violent world, p. 161

Tips for spotting child abuse or neglect, p. 164

Best Practices for Enhancing Children's Social Development

Suggestions for teaching children social skills, pp. 179–192

Suggestions for creating a peaceable classroom and teaching conflict resolution skills, pp. 200–212

Steps for helping an out-of-control child, p. 202

Steps to resolve conflict in the classroom, p. 203

Ways to cope with aggression, pp. 214–218

Best Practices for Supporting Children's Communication Skills

Suggestions for fostering language development, pp. 289–297

Suggestions to foster communication with a child who speaks little or no English, p. 301

Teaching strategies to help children develop literacy, p. 315

Tips for supporting children's literacy development, pp. 315–321

Activities to develop auditory discrimination (Appendix H), pp. 420–421

Best Practices for Curriculum Development

Self-assessment questions every teacher should ask at the end of the day, p. 24

Suggestions for presenting a language-rich group time, pp. 321–324

Management suggestions to help group time go smoothly, pp. 324–329

Tips for using emergent curriculum, pp. 344–352

Best Practices for Supporting Children's Creativity

Tips for encouraging creative thinking through open-ended questions, pp. 294–296

Ways to foster children's creative ideas, pp. 344–349

Tips for "provoking" creative thinking, pp. 349–352

Practical ways to encourage creativity, p. 355

Suggestions for supporting creative expression, pp. 353–356

List of homemade, free and recyclable materials, (Appendix I), pp. 422–425

Best Practices for Supporting Children's Cognitive Development

Tips for teachers on the day-to-day care of young children's brains—Best practices for children's development, p. 7

Developmentally appropriate block play, p. 65

Tips for promoting positive computer experiences for young children, p. 69

Suggestions for deciding what and how to teach in order to support children's cognitive development, pp. 336–341

Suggestions for helping children develop fully at each cognitive stage, pp. 370–372

How to provide opportunities for children to develop concept formation skills, pp. 374–375

Practical suggestions about presenting midlevel thinking and reasoning skills, pp. 376–383

Best Practices in Becoming a Professional

Tips for success in the first few weeks of teaching (Appendix B), pp. 396–399

NAEYC's Code of Ethical Conduct (Appendix C), pp. 400–406

Early childhood journals and newsletters (Appendix D), pp. 407–408

Recommended childhood immunization schedule (Appendix E), pp. 409–411

Summary of communicable diseases and ways to recognize them (Appendix F), pp. 412–416